REBEL

REBEL

My Escape from Saudi Arabia to Freedom

RAHAF MOHAMMED

as told to Sally Armstrong

An Imprint of HarperCollins Publishers

HarperCollins books may be purchased for educational, business, or sales promotional use. For information, please email the Special Markets Department in the U.S. at SPsales@harpercollins.com or in Canada at HCOrder@harpercollins.com.

Ecco® and HarperCollins® are trademarks of HarperCollins Publishers.

Published simultaneously in the United Kingdom in 2022 by William Collins, an imprint of HarperCollins Publishers.

Endpaper map by Mary Rostad

FIRST U.S. AND CANADIAN EDITIONS

Library of Congress Cataloging-in-Publication Data has been applied for.

Library and Archives Canada Cataloguing in Publication information is available upon request.

ISBN 978-0-06-304548-4

ISBN 978-1-4434-6277-8 (Canada hc)

22 23 24 25 26   LSC   10 9 8 7 6 5 4 3 2 1

For all the women who are fighting for their freedom

Contents

REBEL

On the Run

{December 31, 2018}

All that stood between me and freedom was a car ride. For more than a year I'd bided my time, waiting for the right moment to escape. I was eighteen years old and scared to death that my carefully laid plans might backfire. But my heart was full of rebellion against the constant fear, cruel rules and ancient customs that stifle and sometimes kill girls like me in Saudi Arabia. And it soared when I imagined a life away from them.

I had my phone, but my passport was with my eldest brother. Getting it and hiding it so I would have it when the time came to run was key. I was trying to be cool, trying to look like the dutiful daughter packing for a holiday, trying to calm the waves of anxiety as I watched from my bedroom the family prepare for departure and then gather for lunch before setting out for Kuwait.

We were going to Kuwait City, a ten-hour drive from our home in Ha'il, to visit relatives for a one-week family holiday. This was my opportunity to execute my plan. Sitting there watching my brothers carry our suitcases out to the car, I felt a mixture of sadness and excitement. I was torn between hugging my brothers—which is actually forbidden because it's seen as a sexual act—and hoping nothing would get in the way of my decamping.

The bedroom walls around me were bare, with nothing that might make you think a young woman lived in this room. It was not halal—permissible—in this strict society to have signs of life on your bedroom wall. The opposite is haram—that which is forbidden. I remember the teddy bear I had on my bed as a little girl being taken away from me because it was haram—only the Prophet can be imagined in a photo or a form. The drawings I'd once done of people and animals were confiscated, since anything that has a soul is seen as competing with the Prophet and therefore haram. My textbooks and notebooks were scattered around, reminding me that my first semester at the University of Ha'il was over and I would not be returning. I sat on my bed contemplating my life as the Saudi girl who loved her family but could not abide the no-girls-allowed mantra my family swore by; the rebel daughter and sister being driven away by a toxic mix of cultural contradictions.

I was taught in school that Saudi Arabia is the envy of the world; the richest and best country with the most oil; a country that requires its people to make the hajj, a pilgrimage to Mecca, at least once in a lifetime to renew their sense of purpose in the world. Even as a young girl I wondered why oil and resorts and holy treks made this the country everyone else wanted to live in. And it always irked me that a person could make a hajj and be forgiven for everything he does in his whole life, even if he beats his wife or murders a stranger.

My childhood eyes had feasted on other aspects of Saudi: the mountains near our home that beckoned us to come with our picnics and hike to our hearts' content; the vast, ever-changing deserts that never failed to capture my imagination with their undulating sand dunes that changed colour from cool beige to fiery red as the sun rose and set. When my family went to the desert at night, usually to get away from the suffocating summer heat, we would play hide-and-seek in the dark, struggling to get a footing in the soft sand, chasing rabbits and jerboa (a desert rodent) and each other without a care in the world. We ran races and of course the winner got a prize. We'd sing songs, recite poems and dance the traditional dance called Ardah, which is for men but we danced it with our brothers for fun. And always we heard stories from our parents that were different from the ones we heard in school. Some were about the Al Rasheed family who ran this region before the Saud family killed them and took over; others were about the history of our people and the ability of the nomadic Bedouins to subsist in the desert on minimal food and live with simplicity. But the stories we loved best were the ones our parents used to tell us about falling in love, about when they were young. Sharing old stories is like the glue that holds a family together; we never got tired of hearing about the past. I know now we were making precious memories.

From childhood, however, I was aware of the many contradictions in my homeland. While the landscape is mostly shades of beige and white, with patches of green near a water oasis and mountains of outcropping rocks and trees, the softly muted colours of Saudi Arabia are sharply contrasted by the sight of bodies shrouded in black bags moving on the byways. Women and girls over the age of twelve are covered lest a man cast his eyes on their body shapes. In fact, in my family I had to wear an abaya—a loose, shapeless black garment draped over my shoulders and covering my body—at the age of nine, and a niqab, which is like a mask on our faces that exposes only our eyes, at the tender age of twelve. I was a young girl when I began to wonder if this was a form of punishment. If a man can't control himself, why must a woman hide herself behind robes as though it is her fault? And if women do have to be covered, why is it that men who are not in jeans and Western dress wear white robes that deflect the blazing heat, but the women must wear black that absorbs it?

More than half of Saudi Arabia's population of 34 million is under the age of twenty-five, which I felt was a good omen for change. But although the rulers of the kingdom, who claim they act in the name of God, have declared some changes in the strict Islamic rules Saudis live with, and call for tolerance and moderation, they still crucify, behead and torture anyone who doesn't agree with the government. The mutaween—a.k.a. religious police—patrol the streets, even the universities, supposedly making sure the citizens "enjoy good and forbid wrong," which means the shops are closed five times a day during prayers, dress codes for women are strictly enforced and the separation of men and women is fanatically observed, as is the ban on alcohol. In fact, lots of people don't actually pray; girls meet boyfriends in secret places and many drink alcohol without being caught. Since 90 percent of the workforce is made up of foreigners—Saudis don't do blue-collar jobs—if you're sneaking out to meet your friends, the Indian or Afghan man working in the coffee shop isn't going to report you or even understand the language you are speaking. Most of the Saudis who do hold jobs work for the government, where the men nap in the afternoon and tend to gather at about 5 p.m. to socialize until well after midnight.

My family are Sunni Muslims from the Al-Shammari tribe that used to rule the Ha'il region until the Saud tribe took over. Ha'il is the capital of this northwest region. It's the most conservative part of Saudi Arabia, and its people are famed for their generosity, which is why our home is so often open to others who come for coffee or a meal. My family is part of the elite: we live in Salah Aldin, the wealthy part of Ha'il, where there are no shops, only houses, in a big nine-bedroom house with two kitchens (one on the first floor for cooking, the other on the second floor for snacks), ten bathrooms, six sitting rooms and one small garden. We have a cook, a driver and a housekeeper, and there are six family cars; the one waiting for us in the driveway to take us to Kuwait is a black Mercedes. My family also has privileges and a lot of advantages, such as the ability to take holidays in other Arab states like Jordan, Qatar, Bahrain, the United Arab Emirates and Turkey.

But when I think about the feeding of my soul, there is so much we are missing. Consider this: there are no balconies on our house—a good woman would never sit outside where someone can see her. And our windows are closed in case a man might see a woman inside the house. A woman—that is, anyone over the age of nine—can't leave home to visit the neighbours or go to the bazaar, even if only to buy lingerie or makeup, or go out for a walk without a husband, brother or son present to monitor her. We're forbidden to go to the cinema, but we watch American films on our computers. Conversion by Muslims to another religion is illegal. Atheists are designated as terrorists; so are feminists. Homosexuality is punishable by death. Marriage between cousins is the norm; in fact, so many Saudis have married their cousins that genetic counsellors are trying to convince people to stop, as we have dramatic increases in a variety of severe genetic diseases. Having multiple wives is also common, and a man can divorce his wife simply by saying "I divorce you" three times. It's known as "triple talaq."

These are the ingredients of a tribal country that makes its own laws and defies the outside world. This is a country of such hypocrisy that even though religion rules everything—education, the judiciary, the government—95 percent of Mecca's historic buildings, most of them over a thousand years old, have been demolished out of a fanatical fear they will take attention away from the Prophet. Even the ones linked to Muhammad's family have been destroyed. And while most women are covered in black body bags, the female anchors on the television news station owned by the royal family wear Western clothing. It's all for show. Duplicity is the name of the game in Saudi.

Men are everything in my country. They are the decision makers, the power holders, the keepers of the religious and cultural keys. Women, on the other hand, are dismissed, bullied and serve as the objects of men's distorted obsession with purity. It's a complicated and convoluted house of cards that risks collapse in the face of truth-telling.

My father, Mohammed Mutlaq al Qunun, is one of the leaders in Saudi Arabia because he is the governor of Al Sulaimi, a city about 180 kilometres from Ha'il, and interacts in his job with the royal family. He doesn't live with us. He married a second wife, which is legal in Saudi Arabia, when I was four-teen, and took another wife, his third, when I was seventeen. That changed everything for me, my mother and my six siblings. My father stopped coming with us on holidays, and my mother, Lulu, became so depressed, hurt and utterly rejected that even her personality changed. She felt that my father had married other wives because, as she got older, he wanted younger women. And she was right.

That's why this holiday was just my mom and my siblings. I am the fifth child of seven. One older sister, Lamia, is married, and the second eldest, Reem, couldn't come with us this time. So we were six in the car—Majed sat in the front with my older brother Mutlaq, who was driving; Mom and I squeezed in the back with my younger brother, Fahad, and my little sister, Joud. I had to sit in the middle because even though I was wearing the abaya as well as a niqab, I was not to be seen through the car windows. That turned out to be an ideal vantage point for seeing where my brother hid the passports and for carrying out a daring bid to grab mine when he was unaware.

Once we were downstairs and getting into the car, my father turned up to say goodbye and to give each of us money for the holiday. I was already in the car when he arrived. My father has a big warm smile, so engaging that he easily draws people to him. It was a good thing my face was covered with the niqab, because although I was smiling back at him, he would have seen my sadness there as I looked at him for the last time. My feelings about him are so mixed. He treated me very badly and did terrible things to my sister and mother, but somehow I still love him. I felt I was being pushed away by what he and even my mother and certainly my brothers expected of me. They demanded sacrifices I simply could not make. When I cut my hair they locked me up in a room until they figured out an excuse for my shorn look. They finally made me wear a turban to hide my hair and told everyone there'd been an accident and my hair had been burnt and had to be cut. Going outside without my niqab covering my face was an offence that called for severe punishment, and that's what they delivered to me with fists and kicks and slaps. If they were to discover that I had sexual experiences with a man, I knew they would kill me for the sake of honour. Or, at the very least, they would force me to marry a man I didn't know. I had to leave, otherwise I wouldn't be able to live my own life and would have to pay with my life for any mistake I might make. I saw this voyage as the first day of a new life I'd been waiting for ever since I'd begged for the right to attend university in another city and been flatly refused; this was my chance to avoid the trapped lives of my mother and older sisters.

When the car pulled away from the only home I'd ever known, I didn't look back. But as we left the neighbourhood and drove toward the highway, I couldn't help but see the two mountains Aja and Salma off in the distance, symbols of happiness and tragedy that follow me still. Ha'il is sur-rounded by mountains, but these two in the northern part of the city are among the biggest and most recognizable in the region. They are well known to everyone here as the site of a love story. Aja, who belonged to the tribe of the Amalekites, fell in love with Salma, who hailed from another tribe. They declared their love one to the other but their parents refused them permission to marry. Alas, the star-crossed lovers ran away together only to be caught and killed by their families.

Aja was crucified on one mountain and Salma on the other. I knew as a child that this was a love story that was being told as a cautionary tale as much as a story of romance.

The reflection about those long-ago days on the mountains didn't last long, as I was almost immediately consumed with figuring out a way to get my passport. I had watched my brother Mutlaq as he got into the car. I knew he had all our passports—his role as the senior male on this trip was to keep the important documents with him. He often kept our passports in his pocket when we were away because he was afraid they would be stolen, but this time everyone felt at ease since we were all together in the car and going to see family in Kuwait. I didn't take my eyes off him from the time he lowered himself into the driver's seat. Then I saw him slip all the passports into the glove compartment of the car. Apart from the passport, I was also worried that somehow I would lose my phone, that someone would ask to use it to make a call and then keep it. Every single one of my plans was in my phone under a code name, including how I could book a flight anywhere, how I could link to websites, how I could get from Kuwait to Thailand; what to do and where to stay in Thailand; and how to book a flight from there to Australia, which was my planned final destination and where I intended to ask for asylum. The list of my friends all over the world who are also runaways was in my phone as well. I'd been communicating with them for more than a year in Germany, France, the United Kingdom, Canada, Sweden and Australia. I'd received and relied on loads of advice from these friends about how to avoid pitfalls such as Saudi girls arriving in Australia and being asked to call their fathers by officials who don't want immigrants coming into the country. One of my friends alerted me to this, so I arranged with a male friend in the UK to have his name and number with me in case I needed to make that call. I had all kinds of tips for all kinds of potential problems stored in my phone. I also had money, about ten thousand Saudi riyals (US$2,700), stashed away in a friend's bank account. I'd been saving it for about seven months and had the password to the account. My plan was to go to Kuwait with the family and, as soon as I got hold of my passport, escape, get to the airport, buy a ticket to Thailand and connect to Australia. I had friends there who would meet my plane.

It was midnight when we crossed the border into Kuwait. The temperature had dropped to about seven or eight degrees Celsius by the time we arrived at the hotel. I was shivering, but I knew very well it was more from the cold of fear than the night air. It was 2 a.m. by the time we checked into our suite. I still didn't have my passport, as there hadn't been an opportunity to get it. Now I surveyed the hotel suite—two bedrooms (one for my brothers, the other for my sister and mother and me), a bathroom and a sitting room adjoining. I knew this was the place I'd leave from, but having my mother in the same room would create trouble because she's a light sleeper and would wake if I was moving around in the night. So I asked her to sleep in the sitting room. My excuse was that the bedroom was small and had only one big bed for the three of us; she agreed that she'd be better off in the sitting room.

The holiday was nerve-racking. I had to pretend to take part in the shopping and eating and visiting when in fact I was watching and waiting for the best chance to escape. We spent several days shopping at clothing stores in the mall, where I bought a short skirt without any of them knowing and stuffed it into my bag. It was forbidden to wear clothing that showed my legs at home, but I planned to wear it soon in Australia. And having it in my bag was like fuel for the flight from the family I would soon take. We also went to the beach, which was a new experience for me, an experience that hardened my feelings about the sacrifices a woman has to make in Saudi Arabia. My mother told me the women on the beach who were going into the water in bathing suits were bitches—bad girls. I knew they weren't bad. How could it be okay for the boys—my brothers—to be cavorting in the water, swimming, splashing each other, cooling off, having fun, but somehow sinful for me to do the same? I was stuck on the beach wrapped head to toe in my abaya, sweating and swearing I'd buy a bikini when I got to Australia and swim all I wanted. In fact, I don't even know how to swim—girls weren't taught to do anything like that where I come from in Saudi.

Being on that beach was another kind of eye-opening experience. I'd never been to the ocean before, had never seen the tide, with its crashing waves and currents. I was mesmerized by the sight—the incoming tide, the blue colour of the water farther out in the sea and the white caps of the waves as they came closer to the shore. All day long the waves flowed up to the beach and down to the sea. There was something enduring, almost spiritual about the movement, like a ritual on the edge of the ocean. It was such a powerful contrast for me to be wrapped in a false covering and peeking out from behind a disguise while I watched all of this natural splendour.

There was only one day left in our holiday when at last I spied my chance to grab my passport. It was January 4 at two o'clock in the afternoon; my mother, younger sister and I were in the back of the car waiting for my brother to reserve a room for us in the restaurant. The other two boys had gone into the restaurant with him. This was my chance. The front seat was empty. I reached toward the glove compartment of the car and immediately my mom asked, "What do you want?" She couldn't see what I was doing because of the headrest on the seat in front of her. I was still in the middle, but this time I embraced that stifling rule and calmly answered my mother. "I'm trying to charge my phone." I opened the glove compartment, took my passport with my right hand and slid it up the left sleeve of my abaya. Then I brought my right hand back ever so slowly, tilting my left arm up so the passport wouldn't fall out of the sleeve. Once I was sure the passport was safely out of sight, I retracted my left arm into my sleeve until I could get hold of the document and slip it into the small bag I was wearing underneath the abaya. Because the abaya was a flowing garment, no one could tell what I was doing: my dreaded cover became my cover-up. But the act—basically stealing something my brother was entrusted with—had a powerful effect on me. My heart was beating very fast, but I also felt paralyzed; for a moment I couldn't move any part of my body. I could hardly believe what I had just accomplished. I eventually settled back against the seat and sent a text to my friend to say I'd got the passport. I kept writing, "I did it. I did it." But that sense of triumph gave way almost immediately to abject fear that someone in the family would open the glove compartment and see that one of the passports was missing.

By the time we went into the restaurant I was stiff with anxiety, practically like a corpse, unable to talk or laugh with the others. Waiting was an agony. Because we were in a private dining room we could take off our abayas and niqabs. The family could see my face, so I was trying hard to look relaxed. We ordered *machboos*, a family favourite of spiced pieces of chicken with basmati rice, and tea, and I relaxed just a little. But then, while we were eating, I had a nosebleed. I knew it was because I was under a lot of stress, but I didn't want to say anything like that. So when my family asked how I was feeling, I said, "I don't know, I think I'm tired," and then I hoped with all my might that no one would read anything else into the incident. It was bleeding a lot and, to make the situation worse, I was nervous and sweating. I wiped my nose and told the family I would go to the bathroom to clean up, hoping I could buy myself some time to calm down. Once there, my nosebleed got worse and I started to vomit. After a while, when I felt I had control of the nosebleed as well as my own nerves, I went back to my family and tried to act normal; I told them I just felt dizzy and forced myself to join the conversation. When we left the restaurant and got into the car, I tried to keep the conversation going so they would be preoccupied and not think of opening the glove compartment; I kept talking to my brother to divert his attention until we returned to the hotel, where we showered and got ready for dinner with my father's sister.

As we drove to my aunt's house in a town about an hour from Kuwait City, I suggested to everyone that they stay away from caffeine and not stay too late because we would have a long drive back home the next morning. This was our last night away and I didn't want anyone having insomnia; I needed them to sleep early and soundly so I could make my escape.

There were a lot of people at dinner—cousins and friends of my aunt. I looked around and decided with all these people—more than twenty of them, crowded into the house—and with everyone talking and visiting, and with my mother and brothers thinking I was having fun with the girls in the other room, this was my chance to leave. I googled the number for a taxi and texted a cab to come and take me to the airport in two hours. The driver texted back that he couldn't come; I was in an isolated place and taxis didn't come to this area. I was disappointed but not overly worried. Our hotel back in Kuwait City, while not in the centre of the city, was still within the city limits. I figured Kuwait was a big, noisy place, just the sort of place where a young woman could disappear, so I texted the driver again to say if he couldn't come to my aunt's house, he should come to the hotel at 7 a.m. to take me to the airport. By now it was 11 p.m. Although we usually stayed late at parties, I was trying to convince the family that we should go back to the hotel and sleep. At last we said our farewells.

Back at the hotel, I expected everybody to sleep right away, but my mother and brothers stayed in the sitting room chatting. I didn't dare join them, and kept hoping they would go to bed. I called my little sister to play and talk with me but soon enough she fell asleep. The door was ajar, so I could see what was going on in the sitting room. The rest of my family kept talking for three more hours. I was beside myself with worry. This was my last chance. Then, one by one they withdrew— first one brother, then the other and at last my little brother, and then my mother put out the light. My sister and I were alone in the bedroom. She was sound asleep, and soon so was my mother. At 4 a.m. I booked a ticket on Kuwait Airways from Kuwait to Thailand. I knew that once they figured out I'd run away, my parents would track my trip and monitor my bank account with an app that men in Saudi Arabia use to trace their women. The app, which is provided by the Saudi

Ministry of the Interior and downloaded through the Google Play store or Apple App Store, alerts the man if a woman uses her phone, her passport, her credit cards. I knew I had to get rid of the SIM card in my phone and switch airlines once I got to Bangkok so they wouldn't be able to find me. I booked a hotel in Bangkok for three days.

The flight was departing Kuwait at 9 a.m. The taxi was to come at seven. I packed my sister's bag because it was smaller than mine and easier to carry. I stuffed in my toiletries, the short skirt, and my mascara and lingerie too. I used my backpack to take a change of clothes and my papers, money and student card, as well as my passport and some loose cash and bank statements. The room was quiet, the lights in the bedroom very dim; it was still dark outside. When I finished packing, I sat on the bed and looked at my sleeping sister. I wanted to hug her and say goodbye, but of course that would wake her. Instead, I stared down at that dear little girl, memorizing every one of her features before leaving—her long, beautiful eyelashes, the little blue mark on her nose, her soft skin, her lips and her hands. As I listened to her gentle snoring, I was trying to make a mental picture of the way she was sleeping curled up like a baby, her tiny hands under her cheeks.

Joud was only twelve years old. She was so little, so innocent; I feared what they would do to her—the same awful things they did to me. I wanted to remember that sweet face because I knew I wouldn't see it again for a long time. As I looked at her, I wondered if she would hate me for leaving. And I wondered if she would feel hurt that I'd left her. I started to cry and began to hesitate: Should I go and start a new life or should I stay with my little sister? Making that final decision was terribly difficult. But I knew I had to get away, take a chance on what tomorrow would bring. I finished packing, closed the bag quickly and left the rest of my belongings behind. It was time to go. I took the SIM card out of my phone and flushed it down the toilet. Then I put my backpack over my shoulder, lifted the carry-on bag to my chest so it wouldn't make a sound on the floor, and very carefully and quietly tiptoed out of the bedroom and past my mother, sleeping on the sofa in the sitting room. I was trembling with nervousness, but looking at her and hearing her snoring I was sure she was asleep, and I felt more secure. With extreme care I squeezed the door handle and opened the door to the hall. The slight creaking sound it made ratcheted up my fear again, so I decided to leave it ajar in case the noise of it closing might wake her. I left barefoot, with my bag and my shoes in my hands, and ran to the elevator. I could hear voices out in the hall and worried that the sound would now carry into the suite and wake my mother.

At last I was inside the elevator, another step closer to freedom; I slipped my shoes on, and when the elevator stopped at the ground floor, I stepped out and realized I didn't know where the taxi would be, and I couldn't call because the SIM card for my phone was hopefully soaked and ruined in a sewer pipe. I should have kept it until I got to the airport, because now I was stuck with no working phone and couldn't call the taxi driver to make sure he was coming. I tried to act as though I knew exactly where I was going, to avoid having the hotel staff question me. Here was a young woman wandering around at 6:45 a.m. without an abaya. What would they think? Could they stop me? I kept walking toward the back door of the hotel because there seemed to be fewer people there. And then, with all the confidence of a person who used the back entrance every day, I opened the door and walked outside.

I stopped in my tracks, stood absolutely still, as I felt the soft breeze blowing on the back of my bare neck—it felt like a taste of freedom, a freedom I hadn't had since I was nine years old and first told I had to wear a hijab. By the time I was twelve I'd lost the freedom to feel fresh air on my face because I had to wear a niqab. I loved the feeling of the air on my neck and felt like shouting, laughing; the wind touching my face and neck was wonderful, like a spontaneous hug from the world. I felt I could fly at that moment and thought to myself: *This is only the beginning of freedom—the best is yet to come.* I walked along the road behind the hotel to avoid the front entrance because it was on a street with shops and people. I kept walking until I reached the main street. Once there I searched for a coffee shop with Wi-Fi so I could find my bearings and contact the taxi, but there was no coffee shop in sight. Luckily, I ran into a young man and asked him if I could use his phone to call the driver; he loaned me the phone, offered to help carry my bag and waited with me. He asked where I was going. I said Thailand. He then asked, "Where are you from?" I said, "Saudi Arabia." He wondered why I wasn't covered with an abaya and niqab and I said, "My parents are very open-minded."

Finally the taxi came. I asked the driver to take me to the airport, and then connected to the internet through his phone's hotspot. Through messaging apps, I texted my friends; I even phoned them. I didn't feel scared at all. One of my runaway friends who lives in Sydney, Australia, was telling me what to do once I got to the airport. I even did a video chat from the taxi and kept saying, "I did it, I did it." And I took a photo of myself in the taxi and sent it to my friends. The drivers in Kuwait, just like in Saudi, are mostly from India or Afghanistan, so they speak Urdu or Dari. I knew the driver couldn't understand Arabic, so I was free to talk to my girlfriends. I felt victorious.

Upon reaching the airport, I went to the information desk and asked about my flight. The attendant told me I was at the wrong terminal, that this was the domestic terminal and my flight was leaving from the international one. This news threw me, and made me realize I hadn't figured everything out in advance. I asked for the supervisor and explained to him that I didn't know how to get to the other terminal. He saw that I was worried and was very helpful. He told me there was a shuttle bus that went to the international terminal and where to get it, that it was free and that I still had time. I caught the bus and sat there trying to convince myself it would be only a few more minutes and I'd be okay.

On arrival, I stood in line to check in, but when I gave the agent at the counter my passport and my bag, he seemed to be taking a long time to check me in—longer than for the people ahead of me. My heart started racing again. I was scared and asked if there was a problem. He said, "You cannot go." I could hardly believe what I was hearing. My heart sank. I thought there must be an alert about me that he had seen and that this was the end of my life. He must have been calling the authorities; they must have called my father, and now my father would be coming to get me. And my life was over. I tried to get hold of myself and speak with conviction when I asked, "Why can't I go?" He said, "You cannot go to Bangkok because you don't have a return ticket." I tried to convince him to overlook this because I was really going to Sydney after Bangkok, but he said he couldn't. He told me I needed to go to another counter and buy a return ticket to Kuwait, as there is a requirement in Thailand for Saudi citizens to leave after fifteen days unless they have a visa. I went quickly to that counter and said I needed a ticket; the agent told me the fare and the departure times. I was so nervous that I had trouble figuring out the different currency—I was trying to count out in Saudi riyals and put them into Kuwaiti dinars. He seemed to take pity on me and also realized that time was passing and my boarding time was near. "All right," he said, "I'll book it for you. Pay when you reach Bangkok." He gave me a copy of the ticket, told me to pick it up and pay for it in Bangkok, and said I only needed to show this booking to get the visa.

As I left the counter, I felt I'd been extremely lucky so far. Everybody had been nice and wanted to help me; no one suspected I was launching a runaway, the flight of my life. My biggest fear was of being refused or stopped and questioned, of having somebody ask, "Where is your guardian? Where are you going?" I knew this had happened to some Saudi women in foreign airports in Dubai, Egypt and Jordan. But no one had stopped me. The best part was hearing the loudspeaker call the passengers to board the flight. I knew then I had made it; I had gotten out before they knew I was gone. Even if they were in the airport, they couldn't get me now.

Once on board, I sat in the middle seat between two Thai women, and even though the trip took six hours I was wide-eyed: watching the flight attendants, looking out the window at the land I was leaving becoming smaller and smaller. I hadn't slept for a whole day, and I hadn't slept well the previous days of the family holiday, but I didn't want to sleep now.

I was too wound up with excitement. I wanted to savour this moment of freedom. I wanted to contemplate the sky and the morning sun, check out the people on the plane around me and examine the map on the screen in front of me that showed us where we were going. The plane was full, mostly with Thais and a few Kuwaitis. There were young Kuwaiti men in the middle three seats across the aisle from me who asked me why I was going to Bangkok. I said confidently, "I'm going there for fun." One of them gave me his number and said, "Call us. We will meet you and have fun."

I knew the weather would be different when we arrived in Bangkok—hot like summer—so as we neared our destination I went to the washroom and put on a summer top with my jeans. It was the first time I'd uncovered my arms, but I didn't uncover my chest. Clothing protocol was so deeply ingrained in me that, even though I had run away, I kept checking to make sure the top I was wearing was high enough that my chest was covered.

When the plane landed in Bangkok, I was bursting with excitement and anxious to disembark. I followed the passengers inside the airport, not really sure about what to do. All I knew was that I needed to get that visa, and so, when I saw the visa counter, I started walking toward it, thinking how well I was managing everything. Then I saw a man holding a sign with my name on it. My brain was saying *caution, caution, caution*, but he seemed to be very friendly and said, "I'm here to issue you a visa and help you get into Bangkok." I thought this was unusual, but I felt confident that, since I'd made the flight, I was safely away from anyone who could stop me. Despite the clear messages my brain was sending me, I trusted this man who was saying he was going to help me to get a visa. I reckoned the airport office must have sent staff to help visitors. He asked me for all the official documents: the return ticket, my passport and the hotel booking—everything required to issue a visa. I gave all of it to him. He said, "Come with me." We went to a window and he talked to a lady for more than ten minutes. My alarm bells started going off again. The woman looked disconcerted—as if he was telling her to do something she didn't want to do. I badly needed to know what was going on and asked, in the most serious-sounding voice I could muster, for them to speak in English and tell me what they were talking about. They ignored me, and they also stopped talking. A few minutes later I was told I could not enter Bangkok because they could not issue a visa to me. When I repeated that I had all the requirements for obtaining a visa, the woman looked the other way—she would not look at me. I realized the two of them had made some sort of agreement.

And I knew at that moment I had walked into a trap.

Chapter Two

Girl Child

There's an enduring memory that shaped my early years, one of those suspended, dream-like thoughts that comes back to me as I negotiate a new life. It sustains me and makes me believe that innocence and compassion are part of what shaped me. I'm talking about the years before I was old enough to go to school. What comes to me, like a painting with muted colours and soft focus, is a room full of children playing, laughing, singing and teasing. The sound returns to me today like an echo—a faraway melody that makes my breath catch when it plays in my mind. We are together: my two older sisters, Lamia and Reem, and brothers Mutlaq and Majed, and me—five of us—with our nanny called Sarah. It's a flashback I treasure as it pretty much tells the story of the first six years of my life.

The television room on the first floor of our house was where we gathered. It wasn't a huge space for six of us—maybe nine metres square—and the only things in the room were a TV and cushions on the floor, but this is where I spent most of my childhood, and I thought I was the luckiest kid in the world. My sister Joud hadn't been born yet. And my little brother, Fahad, was sickly and had to stay with our mother in her room. I didn't know what was wrong with him at the time, only that he had trouble breathing; he couldn't run, couldn't even walk fast, and he was always out of breath. During those early years, he was never separated from our mother. Later I learned that he had asthma, but when I was very small, I only knew he was sick.

Our nanny, Sarah, was from Indonesia. We always had nannies, though most only stayed two years; they all had kids of their own in Indonesia and came to Saudi to work and make enough money to take care of their own children. But some, like Sarah, stayed longer. I think she was with us for four or five years—certainly during the time I was a child. Sarah was like a mother to us: she was tall, fat and funny. I adored her. She'd make faces: closing one eye, sticking her tongue out and making funny sounds. She'd pretend she was a cat and meow at us, or a dog and bark like a yappy little puppy. She'd tickle us and chase us around the room.

I was an inquisitive and active girl. Sarah encouraged me to seek answers to my questions and stick up for myself as the youngest child in the playroom. She would also wrap her big arms around me if I fell or scraped my knee or quarrelled with one of my brothers or sisters. When the family had parties—and there were many, because that's how Saudi families socialize—we would be with our cousins and aunties, but Sarah was somehow always there like a shadow watching over us. She kept chocolate in her pocket, and like the magical nannies we saw on TV, she would slip us a treat, usually to distract us from whatever calamity we were about to get into.

Until I was seven years old, that TV room was the

centre of my world. We dragged everything in there with us—blankets, pillows, sheets. We built forts and sat inside them, pretending we were the royal princes and princesses or just acting like the kids we were during those happy days. Sometimes at night we would turn off all the lights in the room and hide; one of us would have to catch the others. Even as I describe this story, I find myself holding my breath, remembering how quiet I'd have to be, how I'd sit scrunched up in a corner as still as the air and how we'd stalk each other in the room like leopard cubs until the uproarious shouting of discovery would ring out when one of us was found. We watched TV in that room, cartoons and movies and series from India. Then we'd pretend we were the actors and make our own show. I loved acting; in fact, I believe it was during those shows we did as kids that I decided when I grew up I wanted to be an actor.

Sometimes we slept in that room—all of us, even Sarah; we'd sleep on the floor all together. Sarah would hush us, rub our backs, talk of sleep, and we'd drift off to dreamland. Although we never had a camera and didn't take family photos, the picture I have in my mind of those long-ago nights is of a collection of tired-out children, leaning on each other at odd angles and snuggling together in peaceful sleep.

While we played mostly in that room and told each other stories there, we also went outside into the small garden behind the house some days when it wasn't too hot. There we dug worms and sometimes threatened each other with them; we had mud fights and played hide-and-seek and built more forts and fed our imaginations in what was a fairly small space but seemed like a kingdom to us. We knew every nook and cranny of the yard. We acted like snoops checking to see who or what had been there the night before—gerbils or jirds, which are like mice, but mostly stray cats whose paw prints could be imagined as belonging to wild foxes. And always we produced a show—a play that starred each of us and told a story about five adventurous kids.

There's another memory that feeds my soul when I think about my life in that house with my family. It's the strong, sweet scent of what we call bakhoor. That was the smell of my house. Bakhoor is made from woodchips that have been soaked in fragrant oils such as musk and sandalwood and burned in a traditional incense burner called a mabkhara. The pieces of wood create a rich, thick smoke that billows up and drifts throughout the house. No one wears perfume in Saudi—it's not allowed. But in every house there is the aroma of bakhoor. The smoke wafts onto the walls, the cushions, your clothing and your hair. To me, it is the smell of home. Although it burned in the house throughout my eighteen years at home, today, from afar, the scent of bakhoor brings me back to those years in the TV room and the sense of calm and togetherness created there.

During those early years I don't recall seeing my mother and father very much. Our father had his own bedroom at the far end of our house, with a bathroom and an office attached to it. But usually he was away in Al Sulaimi, where he worked as the governor. There, he lived in a place as big as a palace with a huge garden in front of the entrance, a massive lobby and living rooms for entertaining, and a terrace behind the house that stretched out into another beautifully landscaped enclosure. We stayed there sometimes; there were a dozen other rooms—two kitchens, bedrooms and sitting rooms. But usually he was there without us.

At our home in Ha'il, we saw him on weekends and for special events like parties with our relatives, and also when we went on family holidays or to the mountains in winter or the desert in the summer. All my cousins and friends said the same thing about their fathers—they were pretty much absent from our early lives. As a little girl, I didn't question it. My mom wasn't around very much either. She was a science teacher in a six-room school about a fifteen-minute drive from our house. So our nanny, Sarah, was mother, father and guardian to us.

Then, in 2007, everything changed. It was as though a curtain was drawn over my life—who I was, what I said, the way I could behave. I was seven years old. If I had to put a timeline on the events in my life, I would say this was when I shifted from being a happy-go-lucky child to a girl who no longer understood her place in the family, who wondered why she was being treated so differently, who asked what was wrong with being a girl.

At that tender age, my mother sat me down and told me that the most absolutely prohibited behaviour in the world was for a girl to shout, to be loud, to raise her voice above others. My mother taught me that a woman's voice is like *awra*, which is the word used for the intimate part of a woman's body that she needs to conceal. Some describe awra as the "dark, dirty place of a woman." All of a sudden, my brothers started strutting up to me and raising their fists whenever I yelled or laughed out loud. At that time, I worried more about avoiding their punches than suppressing my laughter. Today, though, I wonder about the effect that suppression must have on kids growing up—laugh and you'll be punished; look downcast and sad and you'll be rewarded for being a good girl.

I was utterly perplexed. I had never been allowed to go outside the front of the house to play. But now I saw my brothers going outside with their friends whenever they wanted, riding bicycles, hanging around in the huge park on our street. There was even a barbecue there, and while the boys cooked food and kibitzed with each other, the girls had to stay in the house. I wanted to have a bike, but my mom was shocked by my request. She said, "Girls don't ride bikes. You would lose your virginity if you were riding a bike, and furthermore, you'd become a tomboy or a lesbian." I used to open the front door just a crack—enough so one eye could peer out onto the street and see what was going on. I'd stand there glued to that opening for four or five minutes at a time, even though getting caught doing that would have meant a harsh punishment for me. Seeing the boys outside having fun made me laugh and smile at first; their good times seemed to be contagious. But then it made me angry—jealous that I couldn't do that too. My brothers played video games indoors; that was also forbidden for girls. I wanted to go to swimming lessons as my brothers did, but I was told that was certainly not for girls. By the time I went to school I knew that boys could play outside, cook in the park, ride bikes and take swimming lessons and that girls were supposed to stay out of sight. I asked my mother why I couldn't play outside. I asked her why older girls and women wore hijab. But she would only say, "If you're a girl, you behave like a girl. Good girls learn to take care of a house for a husband; they wear hijab to show that they are good. They don't do what boys do." They sure didn't. The boys wore jeans and T-shirts; little girls like me wore a *jalabiya*, which is a long, shapeless dress that comes in many different colours and covers your whole body.

We lived on a very quiet street filled with great big houses, but I was never allowed to run over to a neighbour's house to play. My brother used to say that if I was outside alone, someone would rape me. Sometimes I was invited to go to a neighbour's house to play with one of the daughters, but my brother had to walk me there—across the street and down a few houses from our place. When our cousins came, the boys played outside but the girls had to stay inside. When we thought no one was watching we would race around the house hiding in the folds of the curtains and calling out to each other. My mother would say, "Be quiet—I don't want the neighbours to know you are playing." These comments kept reinforcing my feeling that, on the one hand, we were normal kids who liked to play, but on the other, there was something wrong with us so we had to hide our desires. Or was it my mother who was promoting this duplicity?

I wondered sometimes if my mom was different from other moms. She was the only girl in her family, which might have put more pressure on her to conform. But all her friends were married young, while she went away to school to become a teacher and didn't marry until she was twenty-five. Moreover, she was two years older than my dad, while most women were younger than their husbands. It was an arranged marriage, of course—they all are—but my mom told me her parents allowed her to decide whether or not she wanted the marriage they chose for her. She is a very beautiful woman, always has been. And she was meticulous about her appearance; for example, she has one blue eye and one brown eye. She hated that and wore a brown contact lens to cover up the blue eye.

She had plenty of friends, and unlike other women in our family who were not allowed to socialize, my father allowed my mother to go out and visit, so we were with other people a lot. But still the stifling rules for girls prevailed. Whether we were at our house or visiting someone else, women and girls had to sit in the *majlis*—the sitting room—on the floor, perched on what we call cushioned sofas to drink tea and visit each other while boys played outdoors. The lesson for me, as I understood it, was this: a girl should always stay indoors; she should stay calm and never think that playing active games is suitable for her. Although presented as good guidance, it was in fact a strict command that playing is wrong for girls and subservience is right. That was the mantra I began to struggle with even as a seven-year-old. Those customs and traditions stick to me even today, still scratching at my new life, still trying to remind me that people who play and laugh are bad.

There were other customs I began to wonder about as the roles of boys and girls were being defined. At the beginning of my parents' marriage and before my father was working for the royal family as governor of Al Sulaimi, it was my mother who paid the bills. With the money she earned as a teacher, she actually bought our house and took care of the loans for the cars my father bought. There was a stretch of time when I was seven and eight years old that my father was away in Egypt studying. He wasn't away all the time—it was one of those distance-learning schools—but he was gone for weeks, sometimes months. During that period, although we had nannies who basically ran the house, my mother took charge of most things: directing the care of the children, cooking for the family, entertaining the relatives, hiring contractors and cleaners to work in the house. She told us that Dad was away at school because he wanted a university degree that would be useful to the work he was doing as governor.

Most of my friends' mothers did not work outside the home as my mom did. When I was young, the only job women were allowed to have in my region of the country was teaching, because it meant they would work with girls and other women and be separated from men. Still, my mother seemed to me to be a contradiction: she was independent, had her own money and had a career as a teacher, but she was also religious and conservative and forever fussing and worrying about the girls in the family, accusing me of having awra if I dared to laugh out loud, and making sure we were being pious and docile. But at the same time, she was always letting my brothers do whatever they wanted.

I noticed even before I was old enough to go to school that she was conflicted. It wasn't about the difference between a child and her mother—it was something greater than that, as if she had sacrificed the woman she wanted to be and was now insisting that I make the same sacrifice. I felt she was somehow paying a price for her obedience, but as a little girl, I couldn't figure out such nuanced behaviour; I only saw her demands as mean and unfair to me. She didn't let me visit with other children on my street as my brothers did. I wasn't allowed to wear jeans like they did. And I wasn't allowed to talk about what I'd like to be when I grew up, even though I heard the boys talking the way kids do about what they would become when they grew up.

At the age of seven, my world was shrinking—where I played and who I played with defined my whole life. My cousins and I would play "house," arranging the cushions on the floor as if we were mothers with babies visiting each other. We would put a piece of cloth on our heads as though we were wearing hijab and pretend to be mothers. Since having a doll was forbidden because of our religion, we'd scrunch up a pillow and hold it as if it was a baby.

The "no girls allowed" mantra began to seep into my soul like a stain. By now that room that was once full of laughing children had emptied out. Sarah had gone home to Indonesia. Mutlaq and Majed—who were ten and eight years old—had their own room, and they started acting like they didn't want to be with me and like they had the right to boss me around. My sisters Lamia and Reem were thirteen and twelve years old by the time I turned seven and was ready to start school; they also had their own bedroom, and although they watched over me and protected me as big sisters do, they behaved the way girls were supposed to behave—always cleaning the house and learning to cook and doing jobs that pleased my mother. I tried to get out of doing those things. But I felt I had lost my best friends: my brothers and sisters had moved on—without me. I felt very lonely at that young stage of my life. Then my sister Joud was born. Even though she was a baby and had to stay with my mother in her room, I knew there would be a new sister for me to play with. And my little brother, Fahad, who was five, moved out of my mother's room and into my room. He was so special to me, maybe because everything was hard for him or maybe because he was still young and didn't see me as a girl who wasn't worthy of his company the way my older brothers were now treating me. Whatever it was, Fahad became my best friend, and Joud soon became my beloved baby sister.

I remember the day I started school; I was driven there by our driver. All the girls were driven to school. We weren't allowed to walk because there was an unspoken fear that we'd be snatched by men who wanted to have sex with us. I didn't understand what that was about, but I did feel the tension and the fact that somehow I had to be hidden because I was a girl. Some girls saw that as being special. I didn't. I saw it as being denied. And I didn't like it.

But all that aside, going to school felt like embarking on a new adventure that opened all kinds of doors for me—having twenty new acquaintances in the classroom, for example. Even the uniform we were required to wear—a long, pink, shapeless dress and a white blouse—made me feel as though I was part of an important club, a club whose members would become *somebody*. I wore that uniform like a badge of belonging. We brought our lunch to school each day, usually falafels and chocolates, and after lunch we played outside, which to me was a delicious taste of freedom. The tall fence around the school hid us away from prying eyes and allowed us to play games such as hide-and-go-seek. There was a spontaneity about that—running, hiding, calculating my position and trying not to get caught—that thrilled me; it made me feel less confined and it fed my urge to bolt, to be creative and to outsmart the others. But being outside and feeling the wind on my face and the freedom of games of chase wasn't the biggest impetus for me. School was a place to learn and I was thirsty for knowledge. I wanted to know how things worked, why rules were made the way they were, who decided these things. My mother was my teacher for my first two years at school. I felt her watching me, telling everyone I was her daughter, and I liked that a lot. Coming from a family of seven kids meant you didn't get singled out very often, and certainly not if you were a girl. But at school she made me feel special. All the kids at school loved her. I sometimes wondered if she was a liberal person at school and another, more conservative person who thought she had to be tough at home.

Our lessons during the first six years of school included reading, writing, studying the Quran, and my absolute favourite—art. I loved art. I can draw. I'm good at it; everyone knows that. But even with my art there were rules I wondered about. I was never allowed to draw humans. Once, when I was older, I started drawing women's bodies. When my mom found out she was furious—it was as though I'd become a terrorist. Drawing a human was forbidden. Just like the teddy bear that was taken away from me as a child, it's not permitted in Islam. So I drew an eye or a hand; as long as it didn't have a body attached to it, it was not haram. And I drew scenery and food like bananas. And I painted. I escaped into a world of art when I was drawing. Making something beautiful, designing something different gave me an enormous sense of pleasure as well as a boost to my self-esteem, because everyone loved my drawings. I remember one time having an argument with my mom when she refused to buy me a bike. When I started crying, she went out and bought me paints and papers and coloured pencils.

Although I never understood the rule about not drawing people and not owning a teddy bear, I went along with it. Later, when I left Saudi Arabia, I learned there were many other forbidden aspects to our lives that I'd never even heard about. For example, a friend in Canada once asked me how I celebrated my birthday in Saudi. She wondered if boys and girls came together for the party, if we dressed in our best "party dresses and party shoes," if we had cake and ice cream, if we blew out the candles on the cake. She wondered what kinds of presents the kids brought for the birthday girl. And she asked how we sang "Happy Birthday" in Arabic. I was flabbergasted. I'd never heard of such things. We don't have birthday parties, ever! It would be the same as drawing a human being or having a teddy bear—all of it against Islam. I only knew that I was born in the winter. I didn't even know my birthdate until I was nine years old, and that was in the Hijri calendar, which is a lunar calendar of twelve months and 354 or 355 days. As I write, the Gregorian year is 2020; the Hijri year is 1439. I didn't know my birthdate in the Gregorian calendar until I was fourteen; we were travelling and I saw the date on my passport. It's March 11, 2000. It meant a lot to me to know my birthdate, but I'd never heard of birthday parties so didn't consider them.

I think if I could have replaced the word *don't* with *do*, a lot of my childhood would have been very different. For example, the summer when I was eight years old, my father bought a swimming pool. I remember that day as though it was yesterday. It was three o'clock in the afternoon on a sultry, hot day when you could hardly stay outside in the punishing heat. My father was filling the pool with water so we could swim. I was standing next to him, smiling and touching the water, watching the pretty designs it made as I moved my fingers through it. When the pool was full, he smiled at me and reminded me to be careful in the water and then he left. I knew other people had swimming pools, but this was a first for our house. I went inside and I called my brothers and told them the pool was ready for swimming. They raced out of the house enthusiastically and immediately stripped off their clothes and jumped into the pool, swimming in their under-wear. I couldn't decide what I should wear in the pool—my jalabiya didn't seem like a good idea, but I knew I could not take it off. I had nothing else to put on, so I held on to the pool and lifted my leg over the side to get into the water in my dress. Well, you'd think the backyard had been bombed. My brothers started screaming and yelling as though their lives were at risk—not my life, theirs! They insisted I get out of the water and inside the house. They were raising their fists at me and seemed to be out of control. I didn't say anything; I was frightened by their bullying and went inside. I was never close to a swimming pool again.

What stuck with me was my father smiling at me and reminding me to be careful in the water. He would have allowed me to go in the pool. Surely he had more clout than my brothers. But as odd as it sounds, these were issues I would never dare to raise with my father. I always had a feeling that if I'd asked him for a bike, he'd have gotten one for me; if I'd asked him for swimming lessons like my brothers had, I would have had them too. But I never asked. It was one of those unspoken taboos: you just never asked your father. None of us did. I never heard my cousins ask their fathers for anything; it's something that just wasn't done. Maybe that's why I always thought that he would've allowed the things I wanted to do. But I never tested that theory.

Soon enough, my older brothers turned into self-appointed guardians; they began to control me, to check my every move. By the time I turned nine, new rules had seeped into my life and begun erasing the girl I thought I was. I was no longer allowed to sit with my brothers. I could not lie down if my brothers were in the room. I was told not to open my legs ever and to always sit up straight with my legs crossed.

And I learned that I couldn't hug Fahad because it could be interpreted as a sexual act. A girl can kiss her brother on his hair, near the hairline, and a brother can kiss a sister on the forehead. You can kiss your father on his cheek, but now that you're nine you cannot sit on his lap. And you cannot sit outside—ever. You can't open the window, even in your own bedroom. The curtains must be drawn; sunshine never touched the walls of my bedroom. If there's a knock at the door, you cannot answer; you can't say, "Who is there?" No one should hear a girl's voice. I was told you can never walk in public, and if you must work, you can only become a teacher in an all-girls school.

Those are the messages that were delivered to me all day long by my mother and my brothers when I was nine years old. They tried to make me feel that I was less than the boys. I didn't like it and I never believed that it was true, but I didn't know how to defeat it. I never spoke to my older sisters about this as I felt our age difference had created a barrier between us; they seemed to be mature, and to them I was the young, bratty sister. Mostly I felt all alone as a young girl—always thinking, wondering, questioning but not speaking. I worried that there was something wrong with me and wondered why I didn't fit with the others. My life was like a puzzle, but I couldn't put the pieces together at that age. I kept examining the way we were being raised as girls: my mother and brothers would scream at us, punish us, but nothing ever happened to the boys. If I asked why, she would say, "You are a girl; you have to do that," as though I should feel ashamed of questioning her. But I was not ashamed; I thought they were wrong.

It was at about that time that I began to realize I was different from my family in more than my personality. I was the only child with curly hair. I was the only one with brown skin. My sisters had light skin and soft, straight hair. And I was fat, which made me a target for my brothers. They made fun of me and called me nasty names that I try not to remember now. Even my mom made mean remarks about my size. They thought it was funny, but all it did was make me hate my body.

At the age of nine, girls where I lived in Saudi are told that it's time to cover themselves, to start wearing an abaya and hijab. Nobody says why; nobody explains that you have to hide yourself so men won't see you. I saw the abaya as a boring black bag and the hijab as a nuisance that was forever slipping off my head, but I adopted the garb immediately. My family members, who saw me as a rebel, were pleased that I had somehow seen the light and become calm and religious like a good girl, but that wasn't my reasoning at all. I saw the abaya as a way to cover my body, to hide the fat. My mother would even say "Never mind the abaya" sometimes when we were going out, but I was on a mission—determined to cover myself up and stop the cruel teasing.

There were plenty of mixed messages in my young life. For example, we watched movies on television—not American movies (I saw those privately on my computer when I was older) but plenty of love stories and family dramas out of Egypt and India. Although we were not allowed to do any of the things we saw on TV—men being with women, flirting and falling in love—it made me wonder why my mom and dad didn't act like that. I asked my mom why she didn't kiss my dad; she just laughed. They never showed affection to each other. They talked to each other, of course, and shared anecdotes about the events of the day, and they even told us stories. On rare occasions they told us love stories, occasionally about how they met but usually about star-crossed lovers. But I never saw them kiss each other. As far as I know my dad never put his arm around my mom or used terms of endearment like *honey* or *sweetheart* or other words I heard in the movies. Their relationship was something I couldn't figure out.

My father didn't control my mom—she followed the rules for women more because of society's expectations than my father's—but they fought a lot with each other and, to my child's eye, didn't set much of an example of marital love. The one place that true love and unshakable support were abundant was at the home of my cherished grandmother—my mother's mom, who we called Nourah Mom. She was my safe haven through all these years. Although she had more than twenty grandchildren, she always said I was her favourite; we had a wonderful relationship. Sometimes she stayed with us; sometimes I stayed with her. I loved being with her, hearing the tales of her life and taking care of her with all of my heart. She hugged me and listened to my stories and always made a big fuss when I came home from school, as though it made her happy to see me. She fed me whatever she was eating—always with her hands, which was typical of the way older people ate their food. I even slept next to her and remember how she would kiss my forehead as I was dropping off to sleep. If I got up in the night to go to the washroom, she'd come with me and wait at the door just in case I was scared. She seemed to know instinctively when I was upset about something, and I always shared my little-girl woes with her. She understood me and used to say, "You will be a great teacher someday. You will make money; you will do what you want." She made it clear to me that whatever I wanted to do,

I could do. She made me believe in myself. I hung on to her messages as though they were sent by Allah. Even when my mother complained to Nourah Mom about me, my grandmother would say, "Leave Rahaf alone."

One time I took a photo of myself with a camera my brother bought because he was going away to travel and a camera was halal for him. But it was certainly not allowed for me to use it. My mother was horrified, and since she knew I was very close to my Nourah Mom she told her what I had done and expected her to reprimand me. She didn't. Instead, she said to my mother, "Let Rahaf live her own life." I always thought Nourah Mom was open-minded, and I suspected she was trying to set an example for me, to show me that I could also be open-minded if I chose my words and acts carefully. She had benefited from some of the changes for women over the last several decades—girls could go to school starting in 1955 and university as of 1970—but the rules for women had mostly been just as oppressive for her as they were for me. I think she was one of those women who figured out how to play the system, and I felt she was teaching me to do the same. She saw me as different and made me feel that being different was special. My mom made me feel it was bad to be different. I used to think—maybe fantasize—that if my father had spent more time with us, he would think I was different too, would see me the way Nourah Mom saw me and wouldn't allow my brothers to control me. But that fantasy never materialized. Ever since I left Saudi, I have been checking my family's social media posts to see what they say about Nourah Mom; if something happened to her, the family would post it, and all this long distance away, I know I would weep at the thought of not being by her side if she was sick or struggling.

Like me, my friends and cousins were all controlled by their brothers. It's just the way things were. At home, I knew I had to obey whatever my brothers said or I would be punished. They forced me to bring them food, because they said it was my duty as a girl. They even told me how to wear my hair and which words to use while speaking. I had to accept the abuse and the humiliation because there was nothing I could do to stop it.

My brother Majed was the toughest; he had very strong opinions about girls—their appearance and behaviour usually enraged him, and he was never shy about expressing his angry comments. He thought all women were basically bad and that their evilness only needed to be found out. He'd instruct me to take a ribbon out of my hair. When we watched TV he would say things like, "That girl on the television is bad; she's probably cheating on her husband. I bet she smokes and drinks." If an actor was wearing a short dress, he'd be shaking his fist at the TV screen and yelling, "Where is her family?"

Once when I asked my brothers to explain why I had to cover myself and why I couldn't do what they did, the only answer I got was a threat. They told me that if I ever dared to ask these questions in a public place I would be put in jail. They tried to scare me into conforming by saying the mutaween would come and get me. At this stage, I never saw these religious police, but I heard plenty about them: they were the vice-and-virtue squad that went around enforcing the religious rules, such as women covering themselves and men and women staying apart. They claim they do this to protect public morality, and say they have the right to intervene when a Muslim is acting incorrectly. The phrase they use—that I heard a lot when I was growing up—is "enjoining good and forbidding wrong." I always wondered how it was good to hit a girl who was laughing out loud and wrong for a girl to ride a bike. I never saw the mutaween at my school, but I knew that they would give people money if they reported someone who wasn't "enjoining good and forbidding wrong."

For all the criticism I have about the way I was raised, there is a part of my upbringing that was enchanting, just like the blissful days in the TV room with our nanny, Sarah. The day trips we took to the mountains or the desert made a powerful impression on me about how families can be when they are hiking on Aja, the mountain near Ha'il, barbecuing mutton, making delicious salads and devouring sweets. But there was more to those outings: we would pick wildflowers, and since no one was around and we had doffed our abayas, we'd pin the flowers in our hair. It was always winter when we went to the mountains, because the sweltering heat of the summer made it too difficult to hike to where we wanted to go. We'd pick hamadai—little green leaves—to eat. When we were small, our father was with us. He would tell scary stories once it was dark and we'd stay until one or two in the morning. I have such happy memories of those days; I can still smell the wood of the fire and taste the hamadai leaves. My mother had a big jacket called a farwa; she would tuck me in under one arm and Joud and Fahad under the other arm and then sing songs to us. The sky was full of stars, and we often went during a full moon that shone a bright light on our family. My brother Mutlaq carried a gun to keep the family safe from animals that might be drawn to our fire.

And every year we would go on an extended family holiday. We travelled by car—seven kids and two adults crammed in together. The youngest always sat on my mother's lap in the front seat—we didn't have seat belts then—but the rest of us, despite the quarrelling kids do and the stifling rules for girls, rode in the back. We put in ten-hour days sometimes, driving from our home in the north to wherever we were going. And we had fun. We talked all the time, told jokes, played games and sang traditional songs from our tribe, like this one:

عجلوا إلى المجد والسمو
سبحوا خالق السموات
وطن الشجعان وطن الأوطان
للسعودية نحيي روحك
إلى الوطن أنت مخلص دائما
حمل العلم الأخضر

It translates to:

Hasten to glory and supremacy
Glorify the creator of the heavens
The homeland of braves, the homeland of all homelands
To the Saudi, we salute your soul
To the homeland, you are always sincere
Carrying the green flag

And:

انا بدوي من السعودية
أنقذ أعز دولتي.
انا بدوي من السعودية
أنا مشهور باللون الأسود.

Which translates to:

> *I am a Bedouin from Saudi Arabia,*
> *I save my dearest country.*
> *I am a Bedouin from Saudi Arabia,*
> *I am famous for my black colour.*

And:

الله أكبر!

يا بلدي! بلدي ،

عش فخر المسلمين!

يعيش الملك

للعلم. والوطن!

Which translates to:

> *Allahu Akbar!*
> *O my country! My country,*
> *Live as the pride of Muslims!*
> *Long live the king*
> *The homeland! For the flag.*

For some reason that I still don't understand, we didn't act like this with each other when we were at home. Except for those early years when we were together with our nanny, Sarah, in the TV room, we were not close to each other. We didn't share secrets. No one talked about how they felt, what they hated, what they loved. But somehow we had a bond that existed on those holidays. It was a time when I learned the most about my family and their personalities: that Lamia's mood swung a lot; Majed loved to give advice; Fahad was a good listener; and that Joud, when she started talking, never stopped. Reem and Mutlaq were quiet people—they didn't talk much and they loved food—and I learned that my parents were different from each other. My father was usually calm, easygoing, smiling, but my mother was the opposite. She was worried, anxious, always on guard, as though she was the one being judged.

We'd stop the car in the middle of the desert for praying, and then my mother would spread a big cloth on the sand and lay out a picnic lunch that was a feast—the chicken and rice dishes, the sweets and the fruit; every meal on our holiday was like a celebration. My father would tell us stories about each of the towns we drove by, their history and the things that made each place well known: one town had diamonds underground; another was known as a ghost town that looked spooky because no one was there; and another was known for the wolves circling its perimeter. Both parents told us love stories—invariably about two people who fell in love but were killed because their love was a crime, or who couldn't marry because they were from different tribes. These are stories that are usually not shared in Saudi families. I wondered if the car was a safe place for telling them, or whether it was a way of reminding us to be careful with romantic notions. I remember one trip when we were going to Riyadh, the capital of Saudi Arabia, a six-hour drive from home. They told us a story about a prince who raped and killed a woman and then threw her body in the street. We asked my parents if the government would punish him or if he would be spared because he was a prince. We didn't get an answer to that question. Being young, we felt scared because we were going to the capital, where most of the princes live, and we were frightened that something bad would happen to us the same way it happened to that woman.

I knew when I was a child that my family was rich; our house was bigger than my cousins' houses, we travelled more than the others and we had things others didn't have, like bicycles and big TVs. I also knew that my parents were serious about the obligations of Islam—not just keeping girls and boys apart but also doing what we call zakat, which means "charity." My home city of Ha'il is famous for the generosity of its people because it is the home of the poet Hatim al-Tai, who is known throughout history for extraordinarily generous acts. Our parents told us stories about him; he is mentioned in some of the Hadiths of the Prophet and is one of the characters in *One Thousand and One Nights*, the collection of Middle Eastern folktales compiled during the Islamic Golden Age, between 800 CE and 1258 CE. That's why our home was often open to others who would come for coffee or a meal. There were often people eating at our table, people who didn't have enough food themselves. They came to the door—never in groups, only one by one—and my mother would give them clothes, food and money. She'd keep these things wrapped up in bags like presents; I loved being allowed to hand that bag to a stranger. But I wondered why I could answer the door for zakat but not to greet my own relatives.

By the time I was ten, I had my own phone, but as much as the phone belonged to me, it was my brothers who decided if I could use it, who I would call, how long I would talk. They would go through my phone to see if I had called anyone without their permission. Nobody dared to do this. Since being a girl means being tested all the time, your brother will call your cellphone to see if you answer. If you don't, he decides, "Okay, she's a good girl." And if your phone rings, he will answer it and decide if you can talk to the person who has called. I soon learned to hide my phone when my father or brothers were around so that they couldn't interfere with my calls. These rules extended to everything I did. For example, as weird as it sounds, I wasn't allowed to leave the table when I finished my meal until I was told I could go. My brothers left whenever they wanted to, whether or not they had finished their food. They'd run outside to play football, which I couldn't do in case it made me a tomboy or a lesbian!

Later, my mother told me she didn't want me to be around my brothers at all because, she said, I would be a woman soon. I was ten years old when she told me that. Afterwards, she made sure that my brothers and I were always separated and not sitting next to each other. We were siblings; what was she thinking would happen if we sat next to each other?

This is when I learned the meaning of *honour*. My mother sat me down beside her, struck a match and held the flame very near to my body. She said, "Your body will get burned in life and even in the afterlife if you soil your honour or your family's honour." I didn't know what she meant by actions that would "soil" the honour of the family, but I was terrified by this unknown menace called honour, as if it were a monster that was going to devour my soul and torture me.

I remember that day so well. My mother held my head tightly with her two hands and forced me to look downward and said, "This is how you should look—with your eyes cast down at the ground if a man passes by you in public places." Then she squeezed my fingers together and told me that the snakes would eat them up if I moved my fingers between my legs and touched myself. Then she let go of my fingers and moved her face very close to mine and said, in a voice thick with alarm, "You have something that only your husband should take and you will know what that thing is when you grow up." I was utterly perplexed. What was "that thing" I wasn't allowed to know? I wondered if she'd had the same conversation with my older sisters when they were my age, and if Joud, who was only three years old, would be subjected to a lecture like this when she turned ten. I also wondered what my mother had said to my brothers about honour, or if these messages were only for girls. I wondered if boys didn't have to worry about this thing called honour because the girls carried it with them alone.

After that, I was forever being reprimanded, reminded to preserve my honour, to be careful because you could lose it if you sat the wrong way, if you didn't keep your knees tightly together. Our teachers at school reinforced this lesson—that a girl's honour is everything and if it is lost, she might as well be dead. I had a lot of questions, but didn't ask anyone for fear of being singled out as dishonourable. Those were my first lessons about honour, but they wouldn't be my last. My mother said it was because I was about to become a woman. I didn't want to become a woman; I wanted to be a kid.

One day, my paternal uncle came to visit us when my father and brothers weren't home. I saw him coming toward the house and went to open the door to greet him. My mother was literally gasping when she caught up to me and yelled, "Wait! Your uncle is a grown, unmarried man and I won't let you be with him alone." My uncle? What would my father's brother do to me? Why could he not be near me? I had already heard all the speeches about honour and figured it must be connected, so I just backed away and stayed out of sight. Her words didn't surprise me; I had learned that all prohibitions come because of men.

Despite the attempts to keep me in some sort of infantile stage, I became an eleven-year-old with all the meddling curiosity I had carried throughout my childhood. But this turned out to be a year unlike any other, with a tumultuous collection of events that I will never forget. First, we received news that my paternal grandmother had died. I wasn't particularly close to her, so losing her wasn't painful, but seeing my father's reaction to her death was baffling to the point of being shocking for me. I had never seen him cry before. He was enormously sad. I was confused by his show of emotion, something I had never seen in my life. Then, on the heels of that commotion, Joud was diagnosed with diabetes; she was only four years old. Even as I write this so many years later, I relive the agony I felt when I learned that my sweet little innocent sister was sick. I remember crying at the time and fearing that I'd lose her. My mother couldn't stop crying and kept saying that the diabetes would stay with Joud forever. After that, I devoted myself to her, stayed with her, played games with her and made her laugh. Although there was a maid to care for her, I saw myself as her guardian and vowed I would protect her and never let my brothers make her feel the way they made me feel.

But that wasn't all the year had to offer. Months after Joud's diagnosis, my brother Mutlaq made an astonishing discovery. He looked terrified when he came to our mother and said that the closet where my father's weapon was stored was open and the weapon was missing. The whole family clamoured to hear the details. Everyone except Reem. No one could find her. So we immediately presumed she had taken the gun and was going to hurt herself or someone else.

My sister Reem was one of the smartest students in her school; she was absolutely beloved by everybody, and we looked on her as our second mother during our childhood. She took care of everyone. She nurtured us and cooked for us and helped us with our chores. She would spend hours cleaning my father's shoes and washing his socks before he left on a business trip and again when he came back. She was the peacemaker, the one who dried tears and made sure everybody was happy, but she had totally changed that year. She'd become withdrawn, had left the bedroom she shared with Lamia, and preferred to be alone in her own room. She stopped sitting with us at meals and no longer wanted to go to school.

When Mutlaq reported that the weapon had gone missing, we all rushed to her room. I was just behind my mother when she stormed into Reem's bedroom demanding to know where the gun was. It was right there beside her. My mother lunged at the bed and held Reem while Mutlaq grabbed the weapon and left the room with it. I could hardly believe what I saw— Reem was dressed like a man, her suitcase was packed with some of her clothes, and there beside her was a piece of paper detailing her escape plan. She looked scared to death. I had slipped in behind one of the curtains because I wanted to stay, to see what would happen to Reem, and I was certain I would be told to leave. I watched what followed, but from my hiding place I was thinking: *Why does Reem want to run away?* What could have happened to make a fifteen-year-old girl want to leave her home? She looked so scared. She was sobbing and her hands were shaking when she said, "I cannot live in this house any longer." My mother never left her side, but she used her phone to call her brother, my father and my paternal uncle. They all came at once, and with everyone in the room watching, they beat my sister up. They punched her and slapped her, kicked her and knocked her down. It was an almost incomprehensible scene—Reem trying to get away from them, and all of them together snatching her back and hitting her some more. She was bleeding, screaming, begging them to stop. Then, when they thought they had subdued her, she made a bolt for the door and got all the way to the street before they caught her again and pummelled that girl into submission. Her body was beaten but her voice was strong when she ducked another blow from our father and yelled, "You are not my father. You cannot be my father. You know what you did to me. I will not be your daughter ever again." It seemed like she was spewing crazy words. My father said she needed to be taken to a mental institution. I wasn't sure what that meant, but no one left the house that night. I went back to the room I shared with Fahad; we were scared for Reem.

The next day, Reem kept crying all the time. We were told to watch her constantly. All the doors were locked. I understood that Reem had done something wrong by stealing our father's weapon. But what had happened to her? I only knew that Reem, the sister who had helped each of us, needed help now. You could feel the tension in the house, like a taut wire that could snap at any minute. Everyone was whispering, tiptoeing, looking as though there was a powerful secret to keep.

For the next few days, my father spent most of his time at home with my mother, both of them pacing the floor, speaking in whispers. It felt like there was a siege going on. Then my brothers and I came back from school and found my mother sitting by herself, looking very worried. There was no sign of Reem or our father, so we asked her what had happened. She said, "They took Reem to the mental institution." Although I still didn't know what a mental institution was, her words scared me. I hovered around the front hall, waiting for my protective big sister to come home, and when the door opened and she was half carried, half led into the house, my heart broke. She looked like they'd swapped her for another person. Her face was pale, she was dead quiet, and she didn't look at anyone. She kept her eyes downcast and went to her bedroom and fell into a deep sleep right away. People wearing white clothes like doctors followed Reem and my father into the house to give her injections every few hours so she would stay asleep. Of course I asked questions: What is it that they are giving her? Why do they make her sleep all the time? When will she get up and be Reem again? And of course I was told to be quiet, that this was none of my business. I didn't know my sister Reem's secret. I only knew that she was hurting and wasn't feeling safe, and that's why she had tried to run away. And I wondered why.

After that, my mother took up the vigil and sat hour after hour in the living room observing Reem and trying to calm her down when she became agitated. I could tell my mother was scared. I tried to be a good daughter, to stay with her so she wouldn't feel lonely sitting there all day long. I said I would help; I could watch over Reem so my mother could get some rest. I saw this as a chance to get close to my mom. Reem couldn't go out for a visit to our relatives, of course— she was dazed, nervous, unable to manage. The rest of the family went out, but I stayed with my mom to keep her company. Eventually Reem began to feel better, and the three of us sat together, telling each other stories. I liked taking care of them. It made me feel as though I was fitting in, and I wasn't as lonely when I was there with them. I became close to my mom and to Reem and that's what I wanted, because I felt I was helping them and they in turn were helping me.

Nothing stays the same—nothing ever does. Reem became a girl who needed constant care, someone who couldn't manage on her own. I didn't know why, and I wondered about the injections she was getting. Did those drugs somehow dull her memory? Did they make her so dependent on my parents that she couldn't speak for herself? I didn't find answers to those questions. But I do know that as a family we moved on—unchanged. If there was a catastrophe around my big sister, no one learned any lessons from it.

One day when I was almost twelve years old, my older sister Lamia wanted to go and buy some makeup she needed and asked me to go with her. I jumped at the chance. Of course, we had to be accompanied by Majed. I put on my abaya and hijab and got into the car with them. All of a sudden, Majed glanced at me and said, "From now on you won't leave the house without a niqab to cover your face." He raised his voice and spit out that command to me as though he was a royal prince. There was a mixture of arrogance and pride in the way he spoke to me, as if controlling me meant he was somehow a more successful man. In fact, my life was in his hands; his authority was the essence of my being. I had no other option than to accept this reality, so I bought a niqab and put it on. I no sooner slipped the niqab over my head than I felt like I no longer existed. We walked in the busy street, and I felt that

I was invisible; I could see everybody but no one could see me. No one knew whether I was smiling or crying. I wasn't Rahaf but a woman who, like all the women covered in black, was nobody. The girls—or were they women?—walking in front of me couldn't turn around and see my face. I couldn't see their faces. We'd all been erased. It felt hot under that face covering, claustrophobic. I was breathing against a piece of cloth that had started to get damp, making me feel like I was breathing stale air. I wondered if I would suffocate. I was scared that I couldn't breathe but didn't dare tell Majed. Lamia knew instinctively what I was thinking and leaned in close to me and said, "Don't be afraid, you'll get used to it."

There was a lot I was having trouble "getting used to." There seemed to be a game that was played between men and women; it had unwritten rules but everyone knew instinctively how to play. If perchance you didn't know how to play or refused the subterfuge, there were consequences that came in the form of being sidelined. *Ostracized* would be too severe a description, because like the invisible rules of the game, it was never fully obvious that you were not playing by the rules, nor was it always clear that you had been censored.

I had grown up in a family that wasn't open to discussion, that didn't accept each other easily—a family where love was conditional. Now I was supposed to get used to being invisible. I wondered what would come next.

CHAPTER THREE

Holy Orders

By the time I was thirteen and registered at a new school after the summer break in 2012, my life had changed. I was now officially a woman, wrapped up in an abaya fit for an adult—which means it was so big you could fit two of us under it. Not only that, but this abaya had a head covering built into it, so I was hooded as well as robed. Between that and the niqab to cover my face, I looked like some sort of ambulating parcel. There were some perks that came with being a thirteen-year-old. I had moved into a bigger bedroom with my own television (albeit with a restricted number of channels to watch), and I had slightly more control over my own phone. My brothers could still check it whenever they wanted, but I could make my own calls. Best of all, I was now attending intermediate school, which felt like a breathing space away from my family who were at other schools. I had lots of female friends who went to the same school; we knew each other well, but I was the one considered to be mischievous.

This was a time when the emphasis on our role in Saudi society switched gears from "dutiful daughter" to "wife-in-training." The message in the classroom was clear: women were less than men and were created to obey them, care for them and provide them with sexual gratification. I began to examine Islamic laws in my country, laws that allow the state to control people in the name of religion. As a twelve-year-old, I knew that there was no minimum age for marriage; in fact, a man could even marry a five-year-old. And we all knew about the famous camel festivals, held every September in Ta'if, about an eight-hour drive from Ha'il, and about the practice that took place there called akheth—which means "taking"—when girls aged fourteen to sixteen are given as gifts to the elderly members of the monarchy for a few days or weeks. And we knew that child marriage was very common and that most girls in my region were either married or promised in marriage by the age of twelve. It made me wonder what value a girl has other than being somebody's possession.

We started each school day by saying prayers from the Quran and then singing the Saudi national anthem. Immediately after that, while we were still standing, the teachers would check each of us for our appearance—skirts long enough, no short hair (we didn't wear the abaya or the niqab in the classroom). The older girls were checked for makeup and jewellery, which were forbidden. Even the bags we carried had to be plain, without design.

Then the school lessons began, and although we studied geography, history, sewing and housekeeping, religion was the most important subject; its influence affected everything else. Islam is the state religion of Saudi Arabia. The holy book called the Quran is the constitution of the country. Sharia—which means "the way" and comes from the Quran and the

Hadiths, which are the words of the Prophet—are the laws that govern everything from religious rituals to everyday living. And the state police known as the mutaween are the enforcers of those laws.

Saudi Arabia, or "the kingdom," as many call it, is the official home of Islam; the two holiest sites in the Muslim world—Mecca and Medina—are located here, and this is where the Prophet Muhammad lived and died. In fact, the king of Saudi Arabia has an additional title; he is also known as the custodian of two mosques: Al Masjid al-Haram in Mecca and Al Masjid al-Nabawi in Medina. Because the legal system is based on religious law, the leaders of Saudi Arabia govern with what they call divine guidance. That makes the country a theocracy rather than a democracy.

Here is the short version of Islam: it is an Abrahamic, monotheistic religion that teaches that there is only one God—Allah—and that Muhammad is a messenger from Allah. There are more than 1.8 billion followers, which make up 24.1 percent of the world's population. The beliefs of the religion were revealed through prophets, including Adam, Abraham, Moses and Jesus. Like Christianity and Judaism, Islam teaches that there is a final Judgment Day when the righteous are rewarded in paradise and the unrighteous punished in hell. Muslims are Sunni (the majority in Saudi Arabia) or Shia (the minority).

There are five pillars of the religion. They differ between Sunni and Shia, but we are Sunni, so these are the pillars we're taught: shahada, which proclaims the faith and says that there is only one God called Allah, and Muhammad is his messenger; salat, which means prayer, a requirement that is performed five times a day; zakat, which means charity and states that a Muslim must give alms to the poor; sawm, which is fasting done from dawn to dark during the holy month of Ramadan; and hajj, which means making a pilgrimage once in a lifetime to Mecca.

The overriding version of Islam, particularly in my region, is known as Wahhabism, which is an Islamic movement founded by an imam called Muhammad ibn Abd al-Wahhab. The Grand Mufti of Saudi Arabia is a descendent of the founder and has close ties to the royal family. His name—Abdul-Aziz ibn Abdullah Aal Al-Sheikh—is so long it reads more like a URL. An Islamic scholar, he's a scary-looking man based in Mecca and he controls the whole country as far as religion and justice are concerned. I remember that when King Abdullah said he was considering allowing women to vote in the 2015 elections, the Grand Mufti was quick to respond, saying that women's involvement in politics would be "opening the door to evil."

I was still in intermediate school when there was a big fuss made because he went on television to say that the game of chess was haram. Although I didn't know how to play chess and none of the women in my family played, most of the men in our tribe played regularly. It is a very popular game in Saudi. But the Grand Mufti thought it should be otherwise and said, "The game of chess is a waste of time and an opportunity to squander money. It causes enmity and hatred between people." That was a pretty powerful message to the men in the kingdom. Many kept playing but worried that the pastime they loved would soon be forbidden. When I listened to those angry words, as my whole family did, they sounded like another bizarre example of a rule that made no sense. Chess makes you hate your opponent. Really? As a young teenager I knew that was ridiculous and wondered how come others didn't see it the way I did.

When I was a student, we were only taught the Wahhabi version of Islam—a strict, harsh, unforgiving and repressive doctrine driven by coercion and fear. For example, I knew even as a child that it was the leaders of Wahhabism that had destroyed the historic shrines, mausoleums and other Muslim and non-Muslim sites in my country that should have been preserved. Even as a twelve-year-old I thought the education I was getting was mostly religious propaganda that dispersed hate to non-Muslims or anyone who didn't follow Wahhabism.

The thing about studying religion from the very first year of school, as we do, is that we are only taught about one religion, according to one imam. There is no discussion, no questioning—just rote learning and endlessly repeating passages from the Quran. Even though we are all Muslims, we don't learn anything at all about Christians or Jews or Hindus; we only learn about Islam. It wasn't until I was older that I discovered there are many interpretations of Islam, from harshly conservative, like the ones we live with in Ha'il, to more liberal ones elsewhere. I found out that the majority of Sunni and Shia Muslims worldwide disagree with the interpretation of Wahhabism, and many Muslims denounce it; some even claim Wahhabism is a source of global terrorism. But at my school and in my home in Ha'il, we were taught that the only acceptable interpretation of Islam was Wahhabism. All others were considered evil, sinful, even punishable by death.

I sometimes thought Wahhabism should be called the la-la religion, which means "no-no" in English. No, you can't do this. No, you can't do that. The list of rules we were taught at school was withering, especially for young people anxious to learn about the world. No performing or listening to music unless you adhere to strictly prescribed types (like singing the national anthem and, I suppose, like the songs my mother sang to us as children; those must have been halal). Even then, you may mistakenly be drawn into music considered haram. No dancing or fortune-telling, no amulets—those are the small charms Arab people often wear to ward off evil. Only religious television programs were allowed, though we watched TV all the time and it wasn't religious, but I didn't mention that at school. The lists were long: no smoking, playing backgammon, chess or cards, absolutely no drawing of human or animal figures. They're considered graven images, but I'd learned that as a small child when I took up drawing with a passion. No acting in a play (we did that in the TV room with Sarah and no one said anything to us) or writing fiction—both are considered forms of lying. No recorded music played over telephones, no sending of flowers to friends or relatives who are in the hospital. Believe it or not, Wahhabism even says whistling is a sin.

I knew there were many practices that other Muslims did, such as celebrating the Prophet's birthday and using ornamentation at mosques, but our Wahhabi interpretation of the Quran forbade such activities. More no-no, I thought. Women were allowed to drive cars in other countries but not when I was growing up in Saudi Arabia. Dream interpretation was seen as sinful. Our teachers told us that dissecting bodies, even in criminal investigations and for the purposes of medical research, was forbidden. But I wasn't interested in dissecting cadavers at that point in my life, so I struck that no-no off my list as something I didn't have to fret about.

We were also taught to avoid friendships with non-Muslims, and all the cultural practices foreigners have, such as Valentine's Day, Mother's Day, celebrating birthdays or having a dog as a pet. We were even taught not to wish a non-Muslim well on any of their holidays. More no-no. They taught us that Islam forbids women to travel or work outside the home without a husband's permission. Sexual intercourse out of wedlock may be punished with beheading, and gender mixing of men and women is forbidden, as is the mixing of Muslims and non-Muslims. Atheists are viewed as heretics in Islam and can be punished by death.

But as I looked around and asked questions, I found there were invariably exceptions to the rules. My mother was a teacher, but I don't think she had to ask my father for permission to go out and teach school. And at the King Abdullah University of Science and Technology, Muslims and non-Muslims could mix.

It's certainly a no-no for women to break the dress code. Men are required to wear a thobe, which is a long, white robe, and a red-and-white-checkered scarf called a ghutrah, but most men I knew only wore that for special occasions; usually they wore Western-style clothes like pants and a jacket, or jeans. For women there were no exceptions—at age nine, a black abaya that covers every bit of your body except your hands, and by twelve, a niqab that covers your face is absolutely required. To disobey is to beg a harsh punishment.

Some men in Saudi Arabia still think education for girls should be another no-no. There's an expression they use: "Letting a girl go to school is like letting a camel put his nose in the tent—eventually the beast will push his whole body inside and take up all the room in the tent."

In addition to religious instruction, school was also the place to hear bits and pieces about the West—that life in the West was good, that the people there had freedom, that they don't kill girls for doing things that are wrong. I was really shocked when I heard that girls had boyfriends that everyone knew about and that they walked outside together—holding hands.

I learned early to keep a secret—to lie if I must—to avoid punishment, even death. For example, as I was turning twelve years old, something happened to alter my thinking as well as my life. I developed a crush on a girl. At first, I felt something strange was happening to me. My sisters and all the girl cousins in my family loved boys, even though they were always kept apart, but I liked a girl. My feelings were part of me and I couldn't suppress them. I remember the day when we were playing alone in her room. I came close to her and kissed her. She kissed me back, and then we had sex, just like adults. It was the first experience for both of us. After that, I was always thinking about girls, and I couldn't imagine myself with a boy. When I watched TV, I never saw two girls in love with each other. I began to wonder why there were only boys and girls together. It was one question I didn't dare to ask. If I'd been caught with that girl, the consequences would have been grave; everyone at home and at school talked about honour, forever reminding us that we were "clean people."

Honour—having it, keeping it, protecting it or losing it—was the backstory of every single part of our lives. I knew my family would kill to protect their honour, to eliminate shame. All the suffocating restrictions for women and girls, wherever they were—at home or school or visiting relatives or shopping—were for the sake of honour. And that burden was carried by girls and women exclusively. So having sex with a girl would have been seen as the ultimate in scandalous dishonour. If I'd been caught, my family might have killed me, claiming I'd brought shame to the family, or they might have married me off to some old man who would have had total control of me. Thankfully, I wasn't caught that time.

But I developed a new radar for checking my surroundings. After the really ugly incident with my sister Reem, I didn't feel safe at home. I wasn't sure about what had happened to her, but I knew instinctively it was something that could happen to me as well. This foreboding hung on to me like a fog; I could feel it surrounding me, but I could never be clear about whether I was being warned or if I was carrying the emotional consequences of having been a witness to what happened to Reem. I felt there was an unspoken threat and that if I stumbled into a mistake, I might wind up in the mental institution my father had taken Reem to, or worse. I was so frightened, I started having nightmares where my family came into my room, saw suitcases beside the door and pounced on me and took me away. I was screaming in the nightmare. When I woke up, I took all the suitcases out of my room. I even took satchels and big shoulder bags, anything that could possibly be used for travelling. I was absolutely terrified that my parents might be thinking that one day I would run away like Reem had tried to do. At that stage in my life, running away had never entered my mind.

This was a time when I began to be much more aware of a woman's role in society, and I didn't like what I was seeing. As much as I watched and listened, I could not figure out why the men were always being served and the women were always waiting on them; when men came into the house, the women became silent and subservient. The duties and rules I was being taught were humiliating. And I couldn't figure out why this was accepted by everyone. For example, during parties or family gatherings, the women—me included, now that I was thirteen—prepared the food and cooked lunch or dinner for the men and then didn't even eat with them. The men ate what they wanted and left. The women ate the leftovers. It's an accepted habit among most tribes and is still practised. But it's a practice that infuriated me because it reinforced the notion that women are below men, that women don't even have the right to eat the freshly prepared food. And that's not all. Even in a car, women had to always ride in the back. Although when we were little and away on holidays, my mom would sit in the front with my dad, as soon as the boys were old enough, she was relegated to the back seat. Women endure this second-class status even in small, perhaps inconsequential ways. In our majlis, which is the sitting room, there are couches against the wall for the men to sit on and cushions on the floor where the women sit. The messaging touches every aspect of our lives: you are less—less valuable, less important, less useful, easily erased.

When I was a little girl, I thought that the biggest difference between boys and girls was that boys were allowed to play outside and girls weren't, but as I got older I learned that the low status of women was all-inclusive—it affected everything from what we wore and did to even the way we spoke. I remember a day when Fahad was on his iPad and I started teasing him, calling him names the way siblings do. I was telling him he was stupid and weak and easily scared. He and I were always joking with each other. We even swore at each other using words that would get us into trouble if we were caught. The more ridiculous the name-calling the harder we'd laugh; the more gross the swearing the better the insult. Well, on this day my mother overheard us. You would have thought I was threatening Fahad's life. My mother was furious and erupted with a diatribe I've never forgotten, saying I must never ever speak to him that way, that I could injure his alrujula, which means "virility" in English. She said I must never subject a boy to cursing or violence because it makes them weak. Then, as though to make amends with Fahad, she told me that I was a stupid girl, worthless, that I belonged on the street. She took my heart apart with her words that day. I even feel the pain now as I remember the sound of her voice, filled with hatred toward me. I was so angry I shot back, "You don't deserve children and a husband and a big family like us." My reaction was so potent I even felt a grudge toward my little brother, who I adored. I realize now that my grudge was misplaced. I wasn't angry at Fahad; I was angry because my mother was making it perfectly clear that Fahad was more important than me, that she favoured and protected him because he was a boy. She was telling me that girls need to hear words like the ones she seared into my soul because girls need to be broken, that weakness and submission make a girl beautiful. I will never forget that fight. It's carved into my memory like a scar.

At school, as an intermediate schoolgirl, I was hanging around with other teenagers and I thoroughly enjoyed that, but at family gatherings, I liked being with the older women. Most of them were several decades older than me, and they gathered in a separate room from the teenagers. I liked listening to their conversations and learned a lot about our society from their recounting of tales and sharing of secrets and tips for survival. For example, they would talk about how to cope when your husband screams at you: "Look down and be quiet so he will see that you are a good woman." Or what to do when visiting his family: "When he forces you to cook and clean for them, just do it." And when he comes home from work, "Massage him and clean his feet." I absorbed these narratives like foul-tasting medicine. The men don't do any of those things for their wives. There's no pleasure or romance or love in these acts; they are all about service. And to my young eye it was about women fulfilling their role as being stupid and weak because that's what the men want—just what my mother told me I needed to be.

They also talked about sex, mostly about how to make sure your husband didn't take another wife. I'd always thought talk about sex was prohibited, but that wasn't the case among married women. They talked a lot about their sex lives—what was good in bed, what wasn't. Some would laugh and tell the others what they loved about sex; others would complain about having sex when they didn't want to; and still others would quickly tell the complainers to force themselves, to pretend to enjoy it so their husbands wouldn't cheat and get another wife.

As schoolgirls we used to gossip about sex and about an organization called the Obedient Wives Club, which claimed its role was to teach wives how to be submissive to their husbands. I don't think they have members in Saudi, but their book, called *Islamic Sex*, tells wives how to act like "first-class whores" in order to keep husbands from straying. One of their members claims "a man married to a woman who is as good or better than a prostitute in bed has no reason to stray. Rather than allowing him to sin, a woman must do all she can to ensure his desires are met." Some of the gossip reported that the Obedient Wives promoted group sex between a man and all his wives. The book was banned everywhere, but we all knew about it.

I enjoyed the gossip at school as well as the gossip my mother and her friends shared about sex, but it did lead me to examine the contradictions and confusion in our lives. Why was it allowed for married women to enjoy talking and laughing about sex when it was forbidden for girls to even think about it? Not only that, at these family gatherings there was always gossip about this woman or that girl who was bad— bad because she refused to cook daily for her husband, bad because she had a job and travelled to other cities to work, bad for attracting attention to herself and bad because she rejected a marriage arrangement. I remember once, when the conversation turned to the demands women were starting to make about driving, my mother said, "Any woman who wants to drive a car is a whore."

At school we traded these stories, including one about a professor called Kamal Subhi who claimed that allowing women to drive would spell "the end of virginity" in the kingdom. I wondered about those words. How could the act of driving a car take away your virginity? It was always about women and girls being bad and being punished, and never about men and boys doing anything wrong.

Whether we were visiting Nourah Mom or anyone else, we had to be segregated; the boys and girls were separated from each other at our relatives' homes, in our schools, in public places and even in our own homes if someone was visiting. It was our lifestyle, whether we liked it or not, and it taught us to be uncomfortable around boys, even our male cousins. Like most girls in Saudi Arabia, I became scared of my brothers and my father. I knew they had power over me. I didn't like it, but every year as I got older I better understood the consequences of disobeying them or being in their way or even becoming an object of their fury.

This duplicity in terms of men and women (or boys and girls, for that matter), in terms of relationships and the strict rules we had for everything from speaking quietly to acting as though cooking and cleaning were the most pleasing things a girl could do, became more and more perplexing to me. Sometimes the rules seemed to be simply foolish. For example, my cousins came over to our house a lot but we played in different rooms; the girls played together and the boys were in another room. I remember two of our cousins really loved each other; they were nine years old at the time. The boy told his mom he wanted to marry the girl, but of course that didn't get him permission to be in the same room with her. We thought they were sweet together, so we would hide them behind furniture or distract the adults so they wouldn't notice that we'd found a way to get the two nine-year-olds together. We turned the foolish rules into a game to see how cleverly we could trick the adults into thinking we were obeying their silly edicts.

As a child, I was never allowed to go with my friends to public places, but once I was thirteen, it was permissible to go to my relatives' houses as long as I took Fahad and Joud with me. That's how I got myself to exciting places like the souk, which is our marketplace, without my parents knowing. The girls in my relatives' houses were allowed to go out, but we kept our destination secret. That taste of freedom fed my soul as well as my sense of adventure. Just walking in the park or sitting together on a bench made me feel I was part of the world rather than hidden at home. The souk had everything—food and drink and activities and shops. I remember one time, we ate dinner at a restaurant and went to an indoor amusement park and played a game called Kite Flyer that involved a lot of running and screaming and laughing out loud, which are forbidden behaviours for girls in my family. Some old people nearby got mad at us and told us we should not be laughing and shouting in front of men, but we ignored them. It sounds so innocent, but it was so haram that to this day I remember the pleasure of breaking those rules.

Mind you, there were plenty of near misses with our shenanigans. One afternoon while we were carousing around the souk, we saw a friend of Majed's who seemed to look straight at us. I worried he might have recognized Joud, whose face wasn't covered because of her age, and realized he could see me as one of the seven teenage girls alone in the souk. I held my breath until we passed him, knowing that if he'd seen me he would have run to my brother and reported us all as bad girls. Another time, we ran into the religious police, who called us dirty girls because we were outside without a male guardian. In Saudi Arabia calling someone a foreigner is very insulting, so of course the police accused us of looking like foreigners who don't have men at home to properly raise the daughters. They threatened us with being taken to jail if we didn't go home and, to quote them, "take your filthy bodies off the street." The men in the souk also harassed us, grabbing us from behind, pulling my niqab and hijab off my face and head. One bunch of boys brushed by us, pressing their bodies into ours. They called us sluts and invited us out on a one-night stand; they even pulled money from their pockets to say they would pay for time with bad girls. One boy even hit a girl because she was wearing high-heeled shoes.

This is very common behaviour when boys run into unaccompanied girls. It's all about Saudi justice: you broke the rules about being invisible, so you will pay the price of your crime. They feel they have every right to harass us verbally and physically and that we deserve their rudeness and brutality because we dared to be outside the confines of our homes—where we belong. In our society, there's no protection for women and girls from sexual harassment or abuse. In fact, the very opposite situation exists: the draconian regulations that perpetuate a stranglehold on the status of women are rigorously enforced. Women try to seek justice by exposing men who harass them on Twitter—posting a tag that says #افضح_متحرش, which means #exposetheharasser—but invariably there's an onslaught of replies that accuse the woman of causing the behaviour: "You deserve that because you didn't cover yourself," or "It's because you were by yourself," or "If you were at your home that wouldn't happen to you."

All of this is done in the name of religion, which is the real Saudi enforcer: behave the way we tell you to behave or you will be punished by Allah, by the government, by the police, by your family and even by hooligans on the street. It didn't stop me from sneaking out with my friends, because going to public places with my brothers, which was the only acceptable way for me to be outside, was like being a robot—one that couldn't talk or listen or even think. In the souk, I wasn't supposed to talk to the seller; if I had a question, I had to whisper it in my brother's ear and he then asked the seller on my behalf. I wasn't allowed to give the money by hand to the seller or take the bag with whatever I had purchased from him. Even at a medical appointment, when the doctor would ask me questions about why I was there or what was wrong, my father or my brother would answer and explain to him what I was feeling. A conversation would go like this:

DOCTOR: Hello, Rahaf. How do you feel? What's
    happening to you?
MY FATHER: She feels sick in her stomach and she
    vomited this morning.
DOCTOR: How long have you been feeling sick?

If my father didn't know the answer, he would turn and look at me. I would tell him the answer and then he would tell the doctor. This may be hard to believe, but these are the facts of life for girls in Saudi. And if the doctor had to give me a physical examination, a nurse would stay with me while he drew a curtain around me and reached the stethoscope under the curtain, but only under my guardian's observation. Is that not ridiculous?

It was events like this that made me think I was caught in a nightmare. I mean, who can think it is reasonable that a girl cannot tell the doctor her symptoms, that she must tell her father the answer to the doctor's question—which the doctor can hear, of course—and then her father becomes her voice. I used to sit through events like this and wonder to myself, *Who's crazy—them or me?* This kind of inexplicable behaviour in all parts of my life kept telling me that I didn't exist, that I was only on this earth to serve a man, to wash his feet, make his meals and act as his sex slave and baby-making machine.

But why did other girls and women go along with it? How could Lamia, my lovely older sister, think that this was okay? And how about my mother, a strong, seemingly independent woman who had a job and earned her own money—how on earth could she stoop to what I saw as a masquerade? What did the fathers and brothers, the government and the religious leaders think would happen? Did my father really think the doctor would attack me sexually, or did he suppose that a young girl like me would jump into the arms of this middle-aged doctor if he wasn't watching over both of us? No matter where I was—at a doctor's appointment or in a restaurant or a shop—I could not use my voice.

Mind you, I needed to use my voice in self-defence often enough. One day, a student in my school told the teachers that I'd kissed a girl and had relationships with other girls. The story spread around my school like a fire. I knew there would be trouble, I just didn't know what form it would take. I was summoned to the principal's office, and so were the other girls I'd been involved with. We were cross-examined, scolded and called names like "dirty dykes." Then the principal wrote a report claiming that we had admitted our mistake and told us to get out of her sight. Now I became a pariah to all the teachers, who kept taking me to the side and telling me what a disgrace I was and that I must pray to Allah for forgiveness and start living a clean life. While the reaction from the other students wasn't that bad, since most of them also had relationships with other girls, the response from the teachers was ferocious. I was singled out with a few other girls while the others made sure their relationships were kept secret.

It got worse. After the principal and most of the teachers had chastised us, the math teacher made me and the other girls sit in front of the whole class—facing our fellow students—while she yelled at us and portrayed us as homosexuals whose bodies would burn in hell. She didn't stop there. She made us watch a video on her computer about damnation and burning in hell forever. It was the scariest thing I've ever seen in my life, with scenes of people screaming in agony and flames burning them and their bodies turning into ashes. The teacher kept saying, "You will burn like them." I was so distraught I started crying uncontrollably. Even as I tried to stop weeping and trembling and compose myself, I could see that she was enjoying my distress, seeing me scared to death. It was so awful that the images still haunt me, and all these years later, I still quiver with fright when recalling that video.

But I knew very well that the worst was yet to come. The real terror came when the principal told me with a wicked smirk on her face that she had informed my mother, who was waiting for me at home. I wished the school day would never end. I sat in my classes feeling like I might throw up, so nauseated by the anticipation of my mother confronting me that I couldn't speak. As we left at the end of the school day, my friends gathered around and tried to reassure me. They suggested I deny the whole story and tell my mother the teachers had made it up because they didn't like me.

No one was around when I got home, so I went straight to my room and locked the door, hoping for a reprieve, hoping some other problem had occurred in the family and that my mother would forget about me. That was wishful thinking. Soon enough she was at my door, pounding on it as if to smash it down. I opened it up, sobbing. She grabbed me and started choking me, calling me an infidel and a dishonourable daughter. She was wildly angry, pulling my hair, punching me;

she even bit me. I realized that for my own safety I needed to calm her down, but the more I tried to reason with her, the heavier the blows came down on me. Finally, she let me go, but it was only so she could grab things in the room to throw at me. At that point, with her screaming and me bawling, I tried to hide under my bed. She kept shrieking about her reputation, the family's reputation. I realized then that it was all about the knowing, the need to save face. If no one knows about your so-called crime, you may be spared or get off with a light punishment. But if others know, heaven help you. Your life may be the tool used to silence the gossips and rescue the family's reputation. Honour gets in the way of justice. My goal, from under the bed, was to distract her into shifting from action to talking, to beg her forgiveness, to stop her from doing something really crazy. She came to the edge of my bed and hissed, "The end of your life may be near—ask Allah to forgive you." Then she left and took the door key with her. I was so exhausted that when I crawled out from under the bed and lay down, I fell immediately asleep. I woke up with a pillow over my face. My mother was holding the pillow. Was she trying to kill me or frighten me? I couldn't tell, but I knew I was in danger. When I pushed back and started crying, she turned away abruptly and left my room.

I didn't leave my room for the rest of the day, not even for dinner. When morning came and no one spoke to me about going to school, I knew another step had been taken to control my life. My mother announced that she'd had me suspended from school. That began a treacherous guessing game, with my life being the bargaining tool and my mother's fury driving the episodes. One time, while I was going to sleep, she rushed into my room, came to my bed and put the blade of a pair of scissors against my neck, saying, "Wake up, you homosexual." I gasped and threw a pillow at her. Her face was flushed and the rage was spewing from her with such force that I became scared enough to implore her to forgive me. I promised her that I would never make her angry again and would carry the burden of the pain I had caused her for the rest of my life. It seemed to be slightly effective, because she put the scissors down and said flatly, "You're not going to school and you're staying home until a man comes to marry you." I said, "I accept your decision," and felt it was a cheap enough bargain to get me through this period.

I was home for two weeks. My siblings were told that my grades were so poor I would not be returning to school. My mother and I didn't speak. But the single victory for me was that my mother didn't tell my father anything about this transgression of mine. One afternoon when I saw her sitting by herself, I asked if we could talk. By now I was overwhelmed with feelings of rejection, as well as horror and regret, and I desperately wanted everything to go back to how it had been before. I longed for forgiveness from my mother. I sat beside her and admitted my mistake; I cried for her forgiveness. Suddenly, she reached her arms out and held me in a hug. She put my head on her chest and, stroking my hair, said she forgave me because I was young and had made a mistake. Then she announced that I would be going back to school, but it would be a different school, away from the girls who had led me astray. I stayed by her side for the rest of the day. Despite my mother's forgiveness that day, I knew that I had lost her trust, and I knew she had decided I was not what she would call a good girl.

❖ ❖ ❖

The new school was small, only thirteen girls in my class. I made friends quickly and vowed to myself as well as to my mother that I'd steer clear of the girls having homosexual relationships. My mother had become stricter with me and commanded my brothers to keep an eye on me and punish me if I stepped out of line. They did. I prayed as required five times a day, but if I strayed even a few minutes they'd threaten to hit me. I had chores to do in the house, and there was a firm understanding that I'd complete the work or face the consequences. I chose to clean the garden and the patio because this only required splashing some water around and I could be outside in the sunshine.

The scolding, hitting and violence in my life were not unique. Every girl I knew endured similar discipline. Even though my phone felt like my personal property, after the incident at school my mother and brothers would take it from me whenever they wanted and check to see what I was up to. One time, my mother was rifling through my phone and came across a porn movie I had downloaded that featured sex between two girls. Her silence and inaction petrified me. I knew what was coming—she was going to tell my father. He didn't engage in disciplining us unless it was very severe, like the time Reem took his gun and was going to run away. So I figured I was in for the beating of my life. But it didn't happen. He had that look of grave disappointment on his face. Then he took my phone, told me he would decide when I could have it again and left.

Two months later I got my phone back as we were leaving on a family holiday to Dubai. I'd been following all the strict rules, so I was in the good graces of the family. Mutlaq, who had become exceedingly religious and therefore excessively judgmental, decided not to come with us because there were lots of women in Dubai who wore Western clothes, and he felt he must not be exposed to their naked flesh. My mother decided to stay at home to take care of Joud and Reem, who were both having health problems. So that left Lamia, Majed, Fahad and myself to go with our father. I was looking forward to a fabulous holiday. I held my passport for the first time on that trip; that's when I found out my birthdate in the Gregorian calendar. Somehow, I felt empowered knowing something like that about myself.

Dubai was very surprising for me: pubs everywhere, alcohol (which is forbidden in Saudi Arabia) being served and foreign women all over the place in short skirts and very high heels. As much as I was supposed to be disgusted with these women, I was in fact envious. I had never thought of ditching my abaya, but watching those girls in their makeup and pretty clothes with the breeze blowing their hair made me start mulling over the idea. They looked so confident, laughing out loud and calling to each other as if they belonged there and were doing nothing wrong. I watched them through a small slit in my niqab. If they'd happened to look my way, they would barely have seen my eyes. I noticed that my father and brothers had their eyes fixated on these women—on their bodies and their breasts. I wondered what Lamia was thinking when she saw the women. Did she want to toss off the black bag that covered her and show her identity, dress as they dressed? Did she want to show the makeup on her beautiful face? Or was I the only one having these thoughts? I decided I shouldn't ask her. And what about the Emirati women who lived here and were wrapped up, like I was, in abayas and yet saw these foreign women every day? I wondered how they must feel. Did they wonder, *Why her and not me? Who decided she would be free and I would be invisible?*

Apart from the time I spent filling my mental playbook with more evidence of discrimination against girls like me, we had a lot of fun as a family on the holiday. But on the way home I wondered about what my father and brothers were thinking when they looked at the breasts of those women. I saw the looks on their faces when they stared at their bodies; this thinking lingered with me on the drive home. It was a lot to contemplate.

I was trying to figure all this out, and how I fit in this family as well as my place in this world, when my father took a second wife and basically my family imploded.

The news arrived by way of air-splitting shrieks coming from the front of the house a few days after we arrived home from the vacation. We all raced out of our rooms to find out what was going on. My mother had collapsed on the floor. She was crying, beating her fists on the carpet and calling my father bad names. Then she jumped to her feet and tried to hit him. He fended off her blows easily, laughed at her emotional outburst and told her to cool off. We presumed she was indulging in a dramatic reaction to some fight they'd just had. When my father saw that all of us had rushed into the room, he turned on his heel and left, saying, "Calm your crazy mother down." She was still weeping uncontrollably when she looked at us and said, "Your father is taking a second wife. He will marry again soon."

Having several wives—four, to be precise—is allowed in Saudi Arabia because it's part of Islam's Sharia law. It comes from an old interpretation of the Quran written when there were many wars and therefore many widows. The intention was that widows needed protection, the kind of protection that comes from marriage. So apparently the Prophet said men could marry more than one woman, up to four, so that women would not be destitute. The actual verse we were taught in school comes from the Quran. It says, "Marry women of your choice, two or three or four; but if ye fear that ye shall not be able to deal justly [with them], then only one."

The caveat is that the husband must treat them equally and share his wealth with however many wives he has. It may sound good, but it is not. Most of the men in my family have more than one wife. They like it—see it as their right—as though they are somehow serving Islam, which in my opinion is nonsense. I'd say all of the women hate it. Think about it: You marry a person (that someone else has chosen for you), give birth to his children, run his household, and then he chooses someone else. He goes and lives with her and you're supposed to say that's okay. It is not. The second (or third or fourth) wife is invariably younger and brings her own entourage to the marriage—her family and, soon enough, babies that we are supposed to embrace and include as additions to our family.

My sisters and I were outraged and sided with our heartbroken mother, but my brothers believed in a man's religious right to take a second woman and told us we should shut up and obey Sharia law. They attended my father's wedding despite my mother's tears and pleas to side with her. I thought it was selfish of my father to leave my mother and marry another woman, but there were signs that I'd picked up early, such as the way my parents acted toward each other when they were together. There was never a show of emotion—they reacted to each other with respect, but not love. For a long time I thought maybe that was normal for my tribe, the way married people behaved. My mom was shy around my father; she didn't sit beside him. But at the same time, I saw some older women in our family who covered their faces in front of their husbands, so part of me decided that must be normal too. My mother also bought into the rules that I saw as so damaging to women. For example, when women first started advocating for the right to drive a car, she would attack any girl in the family who talked about driving. It was as though she was trying to prove to my father that she was a good wife. Now I started thinking those were early signs that meant there was something wrong with the relationship. But why now? Maybe my father wanted a younger woman—the second wife he chose was twenty-eight years old; my father was forty-five. Or maybe he felt my mother was too focused on Reem, who needed a lot of attention.

What followed were terrific quarrels between our parents and accusations hurled by my mother about what she saw as betrayal. She reminded my father that at the beginning of their marriage she was the one who bought the house and paid off the loans for his cars and put up with him being away when he went to Egypt to study. Now she found out he'd put the house and the cars in his name and she owned nothing. She'd say, "I did everything for you, cooking and entertaining guests even when I was pregnant, and this is how you repay me."

My sisters and I felt a second marriage was the same as my father cheating on our mom. Lots of women speak up against men taking second, third and fourth wives, but they can't change it. Some women in my family had even asked for a divorce when their husbands married another woman, and my mother wanted to divorce my father then. But he wouldn't allow it, and she couldn't get a divorce on her own.

Our family life turned into a tragedy as dramatic as a theatrical production as events unfolded, leaving the players wounded and resentful. My father talked about his new wife in front of us when he came for his weekly visit. He talked about her beauty, even though we thought she wasn't nearly as pretty as our mother, and he talked about taking her shopping and to restaurants, like she was some sort of trophy. He thought he was helping us to warm up to this woman. Clearly, he didn't understand that he was fuelling the fire of hatred we had toward her. My brothers visited them in their new house, which created even more tension at our home. And now that my brothers were the men of our house, they started controlling our mother, claiming she wasn't stable enough to leave the house.

During that time my mother gave me advice I have never forgotten. She told me to always make sure I had my own money, to protect myself so I'd never have to ask my husband for anything. She said I should never trust a husband, because no matter how the marriage begins, he will eventually take another wife, and maybe a third and even a fourth, and then he wouldn't be able to give me any money. She advised me and my sisters to continue our studies and seek financial independence "because it is the sole weapon in the future; men don't benefit women." I believed what she was saying and took her words seriously as I navigated my own life and relationships.

In the meantime, her life was a living lesson about dishonour and infidelity. I was a witness to the breakdown, the sorrow in my mother's eyes. I tried to comfort her when she cried and to understand when she was easily irritated. She became impatient, short-tempered and eventually sank into a depression. Rejection does that. It makes you feel unwanted, undesirable—a used-up old wife who's seen her sons welcome this other woman (which to me was more evidence of the duplicity we lived with) and go off to the house my father bought for her and spend time there with them.

While second, third and fourth marriages is not a topic many women speak of publicly, it's certainly popular among lunatic clerics who use their ranting to reinforce the low status of women. There's one called Abdullah al-Mutlaq who regularly sounds off about the goodness of men and the foolishness of women. He claims that if a woman is angry about her husband taking a second, third and fourth wife, she is actually sinning against Allah, which is a very serious allegation in my country. Moreover, he goes on to say the woman should instead pray for the man and his new wife. Once, on a television program, he said, "We always hear about a woman finding out her husband got married and she goes crazy. She turns psychotic. This, my brothers, is haram." Then he said, as though his holy remarks were useful, "If she bothers you about it just divorce her. Men love peace."

Sentiments like this provide further evidence that a Saudi man controls every aspect of a woman's life from her birth until her death. My father and then my brothers, when they became old enough, were in fact my guardians and had the power to make any decisions about my life. I am forever a child in their eyes. So, it seems, is my mother. Even a professional woman who earns her own money could be humiliated and tossed aside, with no recourse. I had always felt she had escaped that fate. But she hadn't.

I was fourteen years old and confronting some tough truths. I felt vulnerable, lonely. I didn't know what to do. So I turned to Allah. I became religious. And as with everything else in my life, when I embrace something, I do it on a grand scale. I had never taken the religious edicts we lived with seriously because I didn't agree with them. But now I was begging for understanding. I started saying the daily prayers at the five appointed hours, and I would chastise my family members if they were not bowing to the east to pray as well. I was reading the Quran to find answers to my questions about marital love and broken promises and told my siblings that we needed to be close to Allah and to ask his forgiveness for our sins and the sinning of those around us. I started talking to Allah—weeping, in fact—hoping I would find salvation. I admit there was a lot of comfort in confiding my considerable worries to Allah. And I also have to say my family heaped admiration and trust on me once I told them I was a child of Allah. I was so over-the-top religious that my younger sister and brother would even bring their troubles to me and ask me to help them.

I kept this up for months, imploring Allah to change our lives, to make everything right again, but nothing happened. On the contrary, things got worse. My weeping turned to anger, and pretty soon my prayers became a litany of accusatory questions: Why does Allah prefer men over women in life and the afterlife? Why do men have the right to marry four wives when women can marry only one man? Why must women wear black and cover themselves when men can wear whatever they like? Why is sex muharram (*really* forbidden, more so than haram) unless you are a wife? And why are dogs considered impure in the religion? I even raged about how getting a tattoo, plucking my eyebrows or having hair extensions were all makruh (hated) in the religion. What does Allah get from forbidding things that don't harm anybody? I found that these questions had no answers in the Quran. What's more, I wanted to ask Allah about the hatred and ruthlessness that had taken over my family. My invocations turned from confessions to demanding to know how a merciful Allah would allow the children in my family to suffer so much. Soon enough I quit praying and lost hope in what Allah could do.

During my last year in intermediate school, I met a boy from my city via a social media platform. We started chatting online, talking about art and music, and after a while we began speaking on the phone and sharing our thoughts and secrets. I developed strong feelings toward him, mostly because we could talk about things I could not discuss with my family. For example, I loathed the notion of an arranged marriage, of having to marry a man before I got to know him and probably without falling in love with him. I shared my views about traditional marriage (as I had seen it) with this new boyfriend, my fury that this would be my fate, and how truly abhorrent I found the thought of any husband of mine taking another wife. He agreed. We were so in sync that I began to consider if I could actually be attracted to a future husband and feel as relaxed with him as I was beginning to feel with this boy. It was reassuring and heartwarming that we were so totally connected. Although I had to make sure my brothers had no idea who I was on the phone with, even with that concern my heart would sing at the thought of talking to him and I would watch the clock, waiting for the time we could meet on the phone and trade our stories with each other.

Our texting and talking developed into a soulful and intellectual relationship that grew with time. I started to wonder what it would be like to have sex with him. I told him that. I wanted to know what it would be like to give myself to a man that I liked and admired and had such strong feelings for, because I was certain that I would never feel this way about a man picked for me in the arranged marriage that I saw in my future. He was a bit surprised by my suggestion but agreed to come to my room.

Since our house is immense, there are lots of rooms that are never used, so one night I figured out a way to let him in and led him to a dark room where we could be alone. There was no flirting or preamble or teasing. We went straight to it on the floor behind a couch. He entered me. I liked it. Afterwards, when we tiptoed out so as not to be discovered, I saw him to the door and closed it softly behind him. I had no regrets, but I realized that although I'd had relationships with girls and thought I was lesbian, in fact I am bisexual.

I continued seeing him but not at our house. I would sneak out when everyone was sleeping and take the key with me and make sure I was back before my brothers woke up to go to the mosque for morning prayers. One of the problems all the girls who had relationships with boys faced was the issue of blackmail. It was like a recurring nightmare. I know one girl who ran away to France because, after she broke up with the boy she had sex with, he threatened to tell her parents and post compromising photos of them online. In my case, the boy knew everything about me and about my family, even where I lived. So I could have been a target. But he was a nice guy and eventually we drifted away from each other.

As much as I was seemingly unscathed by these shenanigans, I knew I was walking an extremely dangerous line. If anyone found out about my relationship with the boy, death would be my fate. Despite the fact that sex outside of marriage is strictly forbidden by the religion, and therefore Saudi society, we all know it goes on behind closed doors. But everyone has also heard stories about two people facing death because they got caught. It's all about honour—the girl and boy are killed to clear the reputation of her family. If the family is forgiving and decides she needs punishment rather than death, they send her to one of the prisons called Dar Al-Reaya, the worst places on earth.

Sometimes called "care homes," the Dar Al-Reaya house girls and young women between the ages of seven and thirty for crimes such as disobedience—defying the dress code, for example, or an unacceptable sexual orientation, or refusing to marry the man of the family's choice. Everybody knows about these prisons for girls, but hardly anyone talks about them. They're scattered all over Saudi Arabia and are basically a dumping ground for families who claim their daughters have brought them shame—often for an offence that can be hooked to sex, because sexual crimes always get the most attention and the biggest punishment. Most of the girls in these places, though, are actually victims of rape and abuse by male guardians in the family, or they are activists who are demanding change.

There are whispers about what goes on there: nine-year-olds locked away in filthy, tiny cells that are infested with rats, being deprived of meals as a form of punishment, sent to solitary confinement to contemplate their rebellious ways. And then there are the infamous "lashes on Thursdays." Depending on your crime, the judge gives you a minimum of forty lashes—every week. The biggest fear on the inside is that girls may have a sexual relationship, so if they're caught touching each other or even looking at each other, they're immediately labelled as homosexuals, forced to wear a cap, subjected to additional lashings or even murdered.

Getting out of a Dar Al-Reaya depends on your parents. If they want you back, they can get you out, usually with an agreement that you will relinquish any claims you may have made about being violently abused physically or sexually by your father, mother and brothers, as well as a promise to reform—and the menacing threat of tossing you back into that awful place if you disobey again. If your parents don't want you released, you're stuck there until the prison officials marry you off to some horrid man who can't otherwise find a woman to marry, or kill you for committing a crime (like having sex with another inmate), or shuffle you off at age thirty to the women's prison, which is just as bad.

There are lots of truly disturbed girls in these places who need help; they need therapy and get none. There are always fresh reports floating around schools that prove these Dars are cauldrons of the mentally ill, the furiously angry and the rebels, and about girls crying and screaming in their cells, making noises like animals and begging for help. These kinds of rumours run rampant in Saudi, but as schoolgirls we only whispered about these places. No one dares to speak out about them for fear of being arrested by the religious police and hauled away to one of them, or being sent away by a family who wants to get rid of a rebellious daughter.

Chapter Four

Hard Truths

The air of excitement was palpable; it was the last day of intermediate school. The "tomorrow" promise was in the air—the feeling that you have now completed the first steps and the world is going to open up to you because the hallmark experience called high school is about to begin. When I was dropped off at school that day, I was anticipating an event that would mark the end of a confusing and often painful time in my life.

As soon as classes were over, we gathered outside in the garden for the graduation party. We were wearing our school uniforms, but the teachers had placed a ring of flowers around each of our necks. Rows of chairs were set up in front of a makeshift stage. Curtains were draped on either side so it would look like a real theatre. There were tables laden with food and sweet cakes, and decorations in the trees. It was hot and sunny when they opened the gates for the mothers to come in. All eyes were on the arrivals as moms carrying presents and flowers flowed into the garden. There were shrieks of joy as each girl spotted her mom and ran to hug her. Some of the mothers cried with pride as they wrapped their arms around their girls. I kept my eyes glued to the gate—where was she? Probably late, I thought. After all, she was the only one of the mothers who had a job. *She'll be here shortly*, I assured myself. I manoeuvred myself to the back of the group to be sure I could see everyone, just in case my mom had slipped in without me seeing her. The girls were opening their presents—bracelets, earrings, gold rings and makeup to mark this special occasion. When the principal started waving everyone toward the seats, announcing that the commencement ceremony was about to begin, I thought, *Wait a minute, what's the rush? My mom hasn't arrived yet.* And then, like the rumble of distant thunder, my gut began to tell me that my mom wasn't caught in traffic or detained at her job or any of the concocted excuses I'd been running through my mind while my eyes were stuck on the gate, waiting to see her breeze through it. She wasn't coming to my graduation. I felt the sting of tears in my eyes, the growing lump in my throat as the heartsick feeling of abandonment began to overwhelm me. I was the only girl in the class whose mom wasn't there. It felt like a punishment, like an announcement—made in front of the whole class and all the teachers—that I was not a valued daughter, that I was unloved and the child of a woman who was ashamed of me and didn't want to be seen with me. It took all the stubborn willpower I could muster to hold back my tears and find my way to the row of chairs designated for graduates and sit with my classmates.

The certificates were handed out and then the principal began giving out prizes. As the teachers and students from grades seven to nine and all the mothers and sisters looked on, I heard her call my name and announce that I had won a prize for best marks. No one in my family was there to watch me walk up to the podium and thank the principal. No one. I was alone. I stumbled through the reception afterwards, congratulating my classmates and trying to make sure they didn't know how crushed I was or even that I cared that my mother had not turned up. Trying to hide my sorrow was hard.

When I returned home and my mom asked me about the party, I burst into tears and demanded an explanation. I asked her how she could have missed an occasion like this; how did she think it felt for me to be the only girl in the school who had no one at graduation, no one to be proud of her, to show their love for her? She told me she hadn't been able to leave work and immediately hugged me and said, "I will make up for this. We will celebrate together." The next day, she came home with a cake and surprised me with the gift of a new phone. I was easily won over by this attention, but the memory of being the sole girl at graduation without her mom stayed with me like a bruise needing time to disappear.

Two weeks later, my oldest brother came to my bedroom and knocked on the door. When I opened it, he started a speech about me being a graduate of the intermediate school and how I was a mature woman now. It was so unusual for him to say such nice things to me. I thanked him and wondered if this meant we could now be more like a brother and sister rather than a guard and a girl. Then he said, "I know that Mom bought you a new phone for your graduation gift. Give it to me so I can see it and check it." I was accustomed to having my phone checked by my brothers, and I unfailingly erased anything I didn't want them to see. But that day he caught me by surprise. I hadn't removed the conversation I was having with the boy I'd had sex with, and I had pictures of myself in my phone, which is totally haram. I felt a rush of heat to my face and a pressure like goosebumps building in my head. I tried to be calm, knowing he would kill me if he found out what was on my phone, while I frantically tried to come up with a reason why I could not give it to him.

I blurted out, "Wait, I need to charge my phone." That didn't work. He ordered me to give him the phone and reached his hand toward me. I felt trapped—and like an animal that's been cornered, I instinctively attacked. I lunged at him with so much speed and strength that I caught him off balance, even surprising myself, and with the power that comes from abject fear, I managed to push him out of my room and lock the door. I figured he'd smash down the door and knew time was scarce, but I hoped I'd bought myself the minutes it would take to delete everything from my phone. To the sounds of his hollering and vile words—he called me a prostitute—and the heavy thud of his fists on the door, I wiped the offensive material off my phone. Then, to my amazement, there was silence outside in the hall. I crept to the door and listened for his breathing, his footsteps, anything. I was sure it was a trick to make me open the door, that he was hiding out there laying a trap for me. Little did I know he'd gone to the kitchen to fetch a cleaver. Soon enough, he was splintering the wood of my bedroom door. Fortunately for me, my older sister Lamia heard the ruckus and came to see what was going on. She knew immediately that I was in peril, and since our mother was away from the house and our father was out of town, she called my mother's brother, who lives in our neighbourhood, and told him we were in danger. Then she told my brother that he'd better calm down as our uncle was on the way.

A few minutes later the most bizarre exchange took place, with me on one side of the door, and my brother, uncle and sister on the other. My brother told them that I was hiding something on my phone and that I'd attacked him and then locked myself in the bedroom, which was proof of my guilt. Then my mother arrived home and joined the inquisitors at my door. In a voice so soft and sweet I hardly recognized it, I heard my mom say, "Dear Rahaf, don't be afraid. Tell us what is in your phone?" All I said to her was "If I am killed by Mutlaq for not giving him the phone, he will have murdered an innocent person." To be honest, I was so scared, shaking and sweating on the other side of that door, but I knew I needed to hide my fear, so I decided to raise my voice and make my own demands known instead. "From the first day you gave me this phone, I have been under your horrid surveillance." Hoping for even a touch of remorse from them, I continued the diatribe: "It's so painful and embarrassing to me to have you always infringing on my personal belongings." I begged them to consider what it would be like if someone grabbed their phones and checked them—how humiliating that is, how much it shows a lack of trust. No one answered. The quiet was nerve-racking. Then my mom's voice broke the spooky silence. She began to calm everyone, reminding them that there were other things for them to do that day and that this confrontation was over and they should leave. Miraculously, that was the end of it.

Over time, I found the confidence to use my phone not just for social interaction but also to find answers to the questions I had been posing—about the rules in this country, the politics, the treatment of women—and not getting answers for.

The summer holiday was almost over and the long-anticipated high school term was about to begin when my mom and Lamia sat me down to explain that girls in high school are watched by others, who check each girl out to see if she's ready for marriage. They explained that I needed to attract the attention of other mothers, who would see me as a potential bride for their sons. In fact, to have multiple other mothers checking me out would be the goal. That, they told me, requires beauty. They announced that we were going to the salon to make my hair straight and soft and change its colour to a lighter shade of black. From there we went to the esthetician, who applied makeup that supposedly suited me, taught me how to put it on and sent me home with what they called a "beautiful mature look." I loved the attention and liked what they did to my hair, and I also enjoyed the fact that makeup was now halal, but all the while I was wondering what this was about. Even our appearance must change to the standards that this society wants so that we are not who we look like on the outside or who we feel like on the inside. As before, I thought our customs were a sham.

The high school my family had chosen for me was a public school about a five-minute walk away from our house that included all grades—nursery, primary, intermediate and high school. After twenty years as a science teacher, my mother had switched to teaching at the nursery level in 2013. So she was there, as was my youngest sister, who was in the primary level.

I felt a sense of family solidarity when my brother dropped the three of us off on the first day of the new term. Most of my friends from intermediate school had decided to go to this school as well, and we felt lucky that we were all in the same class. There were thirty of us—the teachers called us the "spoiled section" because we all came from the same private intermediate school.

The rules in the public schools were stricter; my girlfriends and I were more rebellious than the other students, but this was a new adventure and I was excited about it. When the day was over, I walked home. My mother and sister had walked the short distance to our house, and I knew that Lamia used to walk home from school, so I didn't think a thing about it. But when I entered the house and started telling my mom and sister about my day, and the new school, there was an odd lack of interest in what I was saying. My mother was preoccupied, as if she was waiting for me to stop talking. Then she said, "Your brother Majed is looking for you." Right away, that flush of fear crawled over my body. I asked what she meant, and my mom said he had called and was angry that he couldn't find me at school when he went to pick me up. I tried to lighten the conversation by saying that he shouldn't worry, that I was home safe and sound. But I could see the anxiety on my mother's face and I knew there was going to be trouble.

All of a sudden my brother was there in the house, screaming my name like a person out of control. I thought it might have something to do with me daring to walk home but I felt I had done nothing wrong, so I went straight up to him, feigning surprise and amazement, and asked what on earth was wrong. He slapped me so hard across the face I almost fell down. Of course, he'd presumed that I'd gone somewhere after school rather than coming home and just assumed I was off with a strange man, fornicating. I tried to explain the five-minute walk, the fact that our mother and sister had walked home, but he wouldn't hear any of it. He punched me in the mouth. I was wearing braces on my teeth then, and the blow was so hard it cut my lips and made me bleed. By the time he was finished with me, I had a black eye and he had a clump of my hair in his hand. Beating me up was not enough: he took all my treasured possessions—the computer, PlayStation, phone, and even the key to my room. My mother witnessed all of this. But she spoke not a word to Majed. To me, this was the ultimate mother-daughter betrayal.

Later she told me that she'd felt sorry for me, that she'd sent a message scolding Majed for what he had done and told him that he was wrong and that walking from the school to our home was not a crime. But ever the mother of sons, she then said that she told him that I wanted to apologize, because I now understood that he didn't want me to walk alone. Ever since my father had moved out, my older brothers had taken charge of the house, even telling my mother what to do and how to behave. I could see she was now trying to make a deal with the devil—taking my side but telling Majed I would apologize. I often wondered why he was so hard on me and thought it was perhaps because he was only eighteen months older and felt the need to exert his right to control me. But with my mother basically allowing him to smash my mouth and make me bleed, and not being willing to interfere when he grabbed my personal belongings and stomped off with them, I figured the only way to get them back was to say I was sorry.

I got my belongings back—but Majed won the round because I had to promise I would wait at the school for him to pick me up every single day. He had no trust in women and girls, especially teenagers. He assumed we were all immoral, even his own mother and sisters. He was always suspicious, thinking we had something to hide, so he kept watching us like he was some sort of secret-service agent. And despite the fact that he was only seventeen years old, he was able to control everyone in the house. Why? Because Majed filled the role our society demanded—he was the man in the house and it was his job to boss us around, mete out punishments and generally become the king of our little fiefdom. An outsider could be entertained by the farce of it all, but living with it was anything but comical. It's a terrifying and oppressive thing, to be guarded by someone who forces you to weave deceptive paths through your life.

During those years, the politics of my country were giving me a wake-up call. We'd already been told that this country was so wonderful it was like paradise. We knew that it was an absolute monarchy, which meant the king had power over the state and the government; in other words, he controlled the constitution and the law—everything—and didn't answer to anyone. I wondered why that was a good thing. I also learned, actually had to memorize, that this is one of the youngest countries in the world, founded in 1932 by the House of Saud—a collection of powerful, tribal men descended from Muhammad bin Saud, who was the founder of the Emirate of Diriyah in 1744 and who also unified many states on the Arabian Peninsula to free them from Ottoman rule.

I wanted to know more about the way we lived, so I started exploring online books and articles that were forbidden and learned that my country ranked 141st out of 149 on the Global Gender Gap Index. We have one of the worst human rights records in the world and the most draconian rules for women. While the king and the princes appear in the newspapers and on television all the time, standing with presidents and prime ministers, the women and girls in this country live in a state of threat all of their lives. What I read was that the kingdom was formed with no attention at all paid to role-modelling the women who went before—the Prophet's wife, for example, who owned her own businesses. Instead it adopted this puritanical form of Sunni Islam called Wahhabism.

Even though those rules are now being challenged by women who are driving reform, progress is agonizingly slow because the laws are not codified and jurisprudence is managed by the personal views and whims of men. Stoning, amputation and lashing are used to punish everything from murder and witchcraft to flirting and robbery. Homosexual acts are punishable by death. An eye for an eye is still practised—the eye of the guilty surgically removed. There's no such thing as a jury trial, and often no lawyer. And the presumption of guilt comes with torture for anyone who doesn't confess.

There was a lot of talk when I was in high school about reform—the country certainly had a long way to go. Even though girls were permitted to attend school and university (unlike some of our mothers and grandmothers), we still had guardians—a father, husband, brother or son who controlled our every move. The kingdom claims to be easing restrictions on women: forced marriage became illegal in 2005; a woman joined the ranks of government ministers in 2009; Saudi women were allowed to compete in the Olympic Games as of 2012; and we gained the right to vote in 2015.

The curriculum at my school included math, biology, chemistry, physics and Islamic jurisprudence, which meant studying the interpretations of the Quran. The interpretations invariably came back to how women should behave. When the teacher said, "Never say no to your husband," I asked, "Why not?" Her response was to stage a play about how to clothe yourself. The good girl was hidden under shapeless garments; the other girl was wearing jeans and a shirt and was described as a wicked girl who was trying to seduce men. I asked why men didn't have to wear hijab or cover themselves. We were never encouraged to examine or analyze anything. The teacher told me to stop overthinking the rules, that Allah knows everything. And she warned me that my kind of questions led to atheism, a crime that would land me in prison. To make her point more directly, she said those who preach atheism are put to death. But I couldn't help myself—the rules simply never fit the reality. Consider that going to the cinema was forbidden, but I was watching Netflix on my computer and gobbling up American films and lifestyles and wishing what I saw there was my life.

My illegal online reading also led me to the House of Saud. The king is Mohammed bin Salman bin Abdulaziz Al Saud. He's worth US$17 billion. There are fifteen thousand members of the royal family, and about two thousand are key players. Their net worth is said to be US$1.4 trillion, which makes them the richest family in the world. Members of the immediate royal family have the title "prince" or "princess"; the lesser members are called "royal highness."

We always gossiped about them at school. Most of the royals have four wives, but they can and do divorce them at will, so the names of the wives change. We'd guess at who was rising in power, who was being dumped or jailed or even killed. And we'd gossip about where the wives were shopping and what they bought. The lifestyle is lavish—marble palaces with multi-million-dollar pieces of art bought at auction; gold-trimmed chairs and yachts; and estates in the country as well as homes in the UK and châteaus in France and smaller palaces in Switzerland and Morocco. But like everything else in my country, these advantages are all for men. Their wives and mothers and sisters and daughters and lovers don't fare as well. The same rules that restricted me, a high school student, restricted them. Saudi women are unable to do anything without permission from a male guardian. These things include, but are not limited to: applying for a passport, travelling abroad, getting married, opening a bank account, starting a business, and getting elective surgery. I read that the guardianship system has created the most gender-unequal country in the Middle East. You can't even use social media without permission. To disobey is to die. Honour killing is part of the religious code.

The Absher app is part of what the men call guardianship. I call it abuse. No wonder we learned to use fake names online and dreamed up elaborate ways to sneak out of the house—or, in my case, to sneak someone in.

I used to see photos of the royal princesses mixing with foreigners at state dinners; while the foreign women looked spectacular in their beautiful gowns, the Saudi women were wearing abayas. They could only strut their own fabulous wardrobes while out of the country. I wondered how it must feel to look but not touch, to want but not have, to admire at a distance. Those princesses didn't dare embrace that lifestyle at home. If they did they could be seen as disobedient, a crime punishable by being sent to the horrid Dar Al-Reaya prison for girls.

Riyadh and Jeddah are very different from Ha'il, where I lived. Women have started wearing colourful abayas and you can hear music coming from some of the cafés. There aren't as many mutaween on the streets. We never went to the movies in Ha'il, but in Riyadh both men and women go to the movies; my friend was there and saw the American movie *Black Panther*.

Still, even in the capital city you wouldn't dare criticize the government or question the lack of human rights. The restaurants are still segregated, but there are rap concerts attended by men and women, as long as they stay three feet apart. Women have more freedom, for sure. Some even own their own businesses. But at the time I was in high school, none of these reforms were happening where I lived.

The summer holidays after my first year in high school were filled with family weddings, neighbourhood and family gatherings, all the usual events. I was spending a lot of time in my room alone, like any other teenager searching for answers, going deeper into social media on my phone, finding new networks and meeting new people online. Although most of them had lived in Saudi Arabia their whole lives, these were Saudis I had never met and who had ideas I hadn't even imagined. There was lots of talk on these internet sites between both boys and girls about drinking, doing drugs, having sex. The girls online were wearing sexy clothes, short skirts, see-through blouses—I could hardly believe what I was seeing. It shocked me, but it answered some of my questions. I was apparently not the only one hiding her sexual practices. I started to wonder how many Saudis were really like the ones I saw online, and did they prefer to hide their behaviour or did they hide their behaviour only because of the rules we lived with? All of it made me think again about the delicate net of deception that weaves its way through our society, and that most people likely do whatever they want as long as they can keep it a secret.

In the process of surfing the net I came across a posting from a person who listed the names of books that are banned in Saudi Arabia as well as in most Middle Eastern countries. The post said some of the authors had been killed for what they wrote because their views opposed the state and the religion. The Saudi government blocks many websites, movies and books on the internet, so I began searching for a way to unblock those sites. When I found it, I clicked on the site immediately. Although my curiosity got the best of me, I admit to feeling some hesitation in delving into these sites. What if my mind changed completely? What if I became like those authors who were expressing the same points of view that I had—sharing my opinions, getting my words banned or being banished myself? As women, we'd been conditioned to think the same way, to never question authority, to abide by the rules of the customs and the religion. Opening that door to new ideas, new thinking, scared me a bit. But I was desperate to free my mind, to find the answers to the questions that had lingered with me since my childhood. So I started downloading the books and the movies that were banned, and exploring the websites the Saudi government blocks.

One of the illegal books I read is called *The Absent Truth* by Farag Foda. He explains that most of the customs we use today are based on historical accounts that relied on oral history. In other words, they are so distorted and so out of step with today that they need to be questioned. I was also very drawn to his comments about the Islamists (those who have extreme views and think Islam should influence politics) using intimidation and distortion as a means of exerting power over the people. He says, "They are seeking political power, not paradise or spiritual salvation."

He even says, "As Muslims, we should not be terrorized by self-appointed representatives of Islam. Islam does not give sanctity to anyone but the Prophet." Foda ends his introduction to *The Absent Truth* by saying that violent Islamists should know that the "future can be made only with pen, not the sword, by work and not by retreat, by reason not by Darwish life, by logic not by bullets, and most important they have to know the truth that has escaped them, namely that they are not alone . . . [in] the community of Muslims."

Then I read that in June 1992, two members of the Islamic Jihad shot Farag Foda dead as he left his office with his fifteen-year-old son. He was forty-seven years old. He was simply expressing his opinion. It made me understand that evil is winning over good.

I remember the day I discovered another illegal book online and read it over and over again: *The Hidden Face of Eve: Women in the Arab World* by Nawal El Saadawi. She writes:

All children who are born healthy and normal feel that they are complete human beings. This, however, is not so for the female child.

From the moment she is born and even before she learns to pronounce words, the way people look at her, the expression in their eyes, and their glances somehow indicate that she was born "incomplete" or "with something missing." From the day of her birth to the moment of death, a question will continue to haunt her: "Why?" Why is it that preference is given to her brother, despite the fact that they are the same, or that she may even be superior to him in many ways, or at least in some aspects?

The first aggression experienced by the female child in society is the feeling that people do not welcome her coming into the world. In some families, and especially in rural areas, this "coldness" may go even further, and become an atmosphere of depression and sadness, or even lead to the punishment of the mother with insults or blows or even divorce.

I felt she was speaking directly to me. She captured my thoughts and feelings, the truth in my life. I read more and began to understand there was something grotesquely wrong with the way I was being raised. She says, "We are all the products of our economic, social and political life and our education." And she says, "There is a backlash against feminism all over the world today because of the revival of religions . . . we have had a global and religious fundamentalist movement."

Another book she wrote, called *Women and Sex*, introduced me to misogyny, a term I had never heard before. Basically, it means the hatred of women. The more I read online, the more I realized that the religion and political tools I was living with were based on the notion that women and girls had to be strictly controlled lest their true selves escape the confines of the abaya and the niqab and upset the selfish sensibilities of the men and the boys. What is honour killing if not misogyny? What is child marriage? What is denying girls the right to speak to men if not pent-up fury about what girls could really do if they expressed their opinions? And how about men having four wives and women having one husband—what's that about if it's not a deep-seated denouncement of women?

Nawal El Saadawi also wrote about female genital mutilation. It's a topic I knew only a little about. For all the terrible things done to girls in my country in the name of religion and custom, it was one hideous affliction that I was spared from. Not all of us—female genital mutilation is practised by some tribes in the south of Saudi. We hardly ever spoke of it, but we knew it was done as a so-called rite of passage for girls; that their external genitalia is cut off with a razor when they're about five years old and then they're told they are now women. If they don't bleed to death from the procedure, or die from shock or infection, they face a lifetime of medical problems caused directly by this barbarous act. Some say it's a religious requirement but it is not. It is illegal in Saudi Arabia, but we all know it happens because of people who think it's the right thing for girls. In fact, at my school one day a girl was saying she wondered what it would be like to kiss a boy and another girl said, "I wish your family did the female genital circumcision to you, dirty girl."

Nawal El Saadawi weaves religion, sex and politics into her books so effectively that I began to see them like a triple whammy: a combination that turns into a single assault on women and girls, and that made me change the way I looked at everything in my life. It also taught me that speaking out and demanding change comes with a price. Saadawi lost her job and was even jailed for her writing. There's a personal story about her that stays with me. While she was in prison, she used an eyebrow pencil and a roll of toilet paper to write of her terrible experience and published the results in another book called *Memoirs from the Women's Prison*. She eventually had to flee her native Egypt for the United States, as her life was in danger. I admired her courage and wondered if I could follow in her footsteps.

I was not alone in my thinking at that time. I had friends through social media who were sharing their opinions online. R was one of those friends. Her dream was to live in Britain with her Lebanese lover, which was a thoroughly impossible plan because of our customs and religion. She fumed about these unfair customs and claimed that because she didn't believe in the existence of Allah the rules should not apply to her. In her other life, R the online rebel was a student in a religious school who dressed in black clothes and was always making Islamic speeches and giving Islamic lectures for young girls. She was the picture of piety. No one in her family or among her friends knew her niqab was really a mask that hid who she was. Again, I thought, *How many people are like R and hiding who they really are? How many are religious fakes? How many are living the life they want?*

At school, I went back to dating girls, as many students did, and broke the solemn promise I had made to my mother to never date girls again. For teenage girls in Saudi Arabia, having sex with girls is not unusual. Sex with a boy or a girl can lead to death via honour killing—but families are less likely to take this drastic action when two girls are involved. They are likely to shame the girls and punish them in other ways.

I had a bad reputation at school because I dated a lot of girls at one time, but I felt no remorse or regret. My friends and I talked openly about prohibited subjects. We were sur-prisingly frank with each other about getting rid of the veil (hijab), attending parties, travelling, having sex. The differ-ence between them and me was they stuck to the customs and religion despite their cravings, but I went after what I wanted and usually found it online.

We had seven classes a day for forty-five minutes each. School rules that focused on our uniform—long skirt and loose-fitting blouse, light makeup, long hair and a black abaya—were often the source of my transgressions. The school principal was tough. When any girl disagreed with a rule or arrived late, she was punished, which meant stand-ing outside for almost an hour in the blazing-hot sun or the freezing cold. I spent my share of time out there with my face to the wall because I kept challenging the rules, kept asking for explanations but never ever getting any.

The problem for me was that for every single one of my altercations or violations, the principal called my mother. This was very embarrassing for her, especially since she was known as a good teacher and a strict disciplinarian. As if that wasn't enough, my older sister Lamia had been at the same school, and the teachers were forever comparing us—the brat and the angel. Lamia never broke the rules or fought with teachers or students, as I did. Although I asked that they stop comparing us because we were different individuals, the teachers always had the same reply: You are from the same household and upbringing; how can you be different? I did not reply. I chose to remain silent. The truth was that Lamia and I not only had different personalities, we had very different thoughts, beliefs and desires. Even as a seven-year-old those differences had made me feel like I was alone, with no one to support me.

Those attempts to question the rules put me on another collision course with my mother. She answered my questions with comments such as, "Your sisters don't ask questions like this. What's wrong with you?" If I asked her for money to buy something, she'd say, "I'll give you the money when you are normal." Then I'd ask my dad and he would just give it to me. But even my friends began to turn away from me when I pressed ahead with questions about love and sex and women and religion; because that was all haram, I think they were afraid I would change their minds. So they steered clear of my constant critiques of the life we were living.

Nawal El Saadawi spoke the words I felt. "As a girl there was something wrong in the world around me, in my family, school, in the streets. I also felt there was something wrong with the way society treated me." And she admits being furious when her grandmother said, "A boy is worth fifteen girls at least because girls are a blight." I related to every word she wrote.

My research also showed me that no other country restricts the freedom of women to travel, or to get a passport, more than Saudi Arabia. And that having a male guardian who controls everything from where you go to who you marry is not the law in other countries. It certainly makes violence against women a national sport in my country. And having a medical appointment is a farce, since your father has to do all the talking on your behalf.

What's more, I discovered that because Sharia law rules everything in our country, we have no such thing as family law, so a woman's right to divorce is more restricted than a man's. Divorce is fairly rare except among the royal family, where it's common, but for the rest of us it's more complicated and much more expensive for a woman to get a divorce. She has to pay back the dowry she received for the marriage and prove that she's been mistreated—a pretty tough requirement when men are in charge of everything. Not only that, a woman has no right to be the legal guardian of her children. She may start out with custody, but at the age of seven girls are transferred to their father's custody and at the age of nine boys can decide which parent they want to live with. And when it comes to inheritance, a woman is entitled to only half of what a male heir inherits.

The rules are never-ending. A woman from my district can't study abroad without a guardian tagging along with her. If she's in prison she can only be released to a male guardian after serving her term, and if he decides he doesn't want her, she doesn't get out.

In February 2016, while I was surfing social media, I came across an account on Path (an app we use like Facebook) that caught my attention. The girl's photo looked familiar, she lived in the same city I lived in, and although I didn't know her, I was immediately interested in her. I clicked on her photo and waited for her to accept my friend request. A day later she did. The back-and-forth conversation that followed filled in the details of her life and mine, and I discovered we were both at the same school. We exchanged phone numbers and started chatting. After a while, I started hinting to her about sexual relationships because I wanted to know how she felt about a lesbian relationship. I wasn't suggesting that she become my girlfriend at that point; I just wanted to know her thoughts. Well, she said she did lean toward girls rather than boys, but she also made it clear that her behaviour in public was the opposite to that. I got to know her at school and discovered she was shy and quiet and not one to break the rules. We kept up the conversation throughout the term as friends, not lovers. In July 2016 she told me she wanted to be my girlfriend. I already had a girlfriend, but there was something about my new friend that made me want to keep this relationship special—I didn't want to let her go. Despite the many homosexual relationships I'd had, there was always the unhappy truth that eventually I would have to marry according to the customs and forget the girls I had dated in the past. This time was different. I began to wonder how I could skirt the demands of my country and the laws of this land.

I was preoccupied with those thoughts when the school year ended. Once at home for summer holidays, I turned my attention to my mother, who seemed to have shed her domineering side and become pathetically vulnerable after my father left us to live with his second wife. She was even trying to alter my behaviour at school by way of spoiling me with gifts and treats and affection. For the first time, I could see the look of embarrassment on her face when the school called to report that I'd had fight with a teacher or a student, or that I'd skipped a class or broken a rule (like pulling up my skirt to make it look shorter). I knew she had to apologize to everyone on my behalf, and seeing her shame because of me hurt me. I understood my behaviour was typical of teenagers acting out, but I began to realize that the price my mother was paying for my actions was too high given the unhappy position she'd been in since my father had broken her heart.

That summer, our family trip was to Mecca to do the hajj. This is a religious pilgrimage dictated by the Prophet that every able-bodied Muslim who can afford to is obligated to make at least once in their lifetime. It takes place from the eighth through the twelfth day of the last month of the Islamic year. *Umrah*, which is a short version of hajj, can be performed at any time of the year and can be done in a few hours as opposed to a few days. Whether hajj or umrah, the point of the pilgrimage is to prove that you submit yourself to God—Allah—and to show solidarity among people of the Muslim faith and cleanse your soul of the sins you committed in the past.

Despite the fact that by now I knew I was an atheist and lacking any level of faith, I felt I couldn't refuse the trip with the family. The rituals once we reached Mecca included washing our bodies, cutting a piece of our hair and nails, and entering the holy mosque Al Masjid al-Haram barefoot. I'd done this before, but now, as a non-believer, it felt very strange to realize the mind could be both blindly believing and brilliantly analytical when viewing the black cubic box known as the kabba, which is Allah's home. You need to suspend disbelief to be holy. Millions of people from around the world come here to touch and kiss this black box. We started the required walk—seven times around the kabba. This represents the story of how the wife of the Prophet walked a path among the mountains seven times in scorching heat in search of water for her baby son while her husband was away serving Allah. The Hadith says that as the baby Ismail cried and rubbed his feet on the ground, water began to flow from the mountain—and will flow until Judgment Day. It takes about three hours, walking and praying, to complete the circuit. I saw it as good exercise.

Afterwards, we rested at the hotel in Mecca and then continued our journey to Jeddah to enjoy the rest of the vacation.

At the end of the holiday I convinced my family that we should have a cat. We'd never had a pet before, and although dogs are not allowed (they are haram), I figured I could talk them into a cat. My father was visiting when I raised the idea and he agreed to go with me and Lamia to the animal shelter. I think Lamia went along to supervise, worried that I would choose a cat that was too big or too scary or too wild. We had no sooner entered the shop than all three of us spied the same one—a beautiful, quiet, gentle cat with silky caramel-coloured hair. I named her Sasha. Everyone fell in love with that cat.

There'd been a Twitter campaign that summer that called for the government to "Drop the Parental Guardian Rule." The action gave birth to a slogan—#IAmMyOwnGuardian— which was very popular among women in Saudi. I didn't sign on at first, but watched and listened to what they were demanding. More than 2500 women sent petitions to the king demanding that guardianship end. The petition itself received 14,682 signatures on Twitter. The clarion call was that "women should be treated as full citizens." I was scared to sign in case my brothers saw my name, but I really wanted to be part of this, so finally, during the third week of the campaign, I used a pseudonym and a different email account and added my vote. I felt a flood of empowerment as I hit send because I was at last doing something to change the unfair rules I hated in my country.

❖ ❖ ❖

Like lots of young Saudi girls, my girlfriend was away with her family during those summer holidays; she was studying, but I knew she'd be back in time for the start of the school year. My relationship with her was different than with the others. I could relax with her; she never judged me and always accepted me for who I was. Like me, she confessed that she didn't like to pray, and told me she would only act as though she was praying in front of her family. I did the same. I felt hesitant to tell her that I was an atheist, but when I asked her if she believed in the presence of God, she said she didn't really believe in religion. I started sending her information I found online. We had lots of discussions about all the prayers and requirements we'd been taught and agreed finally that it wasn't something that either of us could believe.

I was becoming more and more dependent on this relationship. I didn't want to lose her, but I wasn't fully committed to her. Having someone in my life who knew my truth and still liked me was a gift I'd never received before. I didn't want to tell her that I had thoughts about living outside of Saudi Arabia, but she'd been to North America and I wanted to know what she'd learned—what it was like to live there. I asked about the laws and the lifestyles. She showed me her photos and let me read her diary. Despite being in North America she was forced by her family to wear her abaya and the niqab and wasn't ever allowed outside without them.

Although Saudi Arabia claims racism is against the law, there's an abundance of it everywhere—in the workplace, in the shops, even at school. Anti-Semitism is common even in the media. As for Black Arabs and Africans who work in Saudi, being subjected to derogatory language and outright mockery is common. There's obvious contempt for anyone who is Black or African. Television comedy series use blackface at will and depict Black people as lazy, stupid and engaging in sorcery. There is a lot of judgment in Saudi around who your family is, what your skin colour is, how you wear your hair. Many of my classmates were guilty of racial and religious slurs. It left me wondering who I would be if I didn't match their description of "good" and "beautiful" and "successful." One day in February 2017, I was showering and looking in the mirror and wondering who I would be if I cut my long hair off. Would my family, my classmates, my society still like me if I had short hair? Without giving it another thought, I picked up a pair of shears and started cutting. As long swaths of hair fell into the sink, I got bolder and cut more and more and more until my hair was short. I felt a surge of freedom. *This is me! This is who I want to be. I am not a superficial example of my society. I am Rahaf!*

I have to admit that the new look was shocking—and that I hadn't done a very good job. I convinced Fahad to act as my guardian and come with me to the hair salon where I had my hair cut even shorter to make it even. I knew I needed to face the others, but since it was the cold weather months, I decided to wear a hat that covered my head entirely

My family was caught up with the preparations for Lamia's wedding at that time, so I escaped their scrutiny for a while. She was marrying the son of a family in our city; my brothers were friends of his younger brothers. Lamia had never met him but agreed to the marriage, which was arranged by our father. The two families went to a popular estate outside of Ha'il called The Farm so we could have a gathering and get to know each other. We were staying there overnight; I wanted to tell Lamia that I'd cut my hair but hadn't found a chance to do so. Here at this retreat I called her to my room and told her what I had done. Even though my words were plain—"I cut my hair"—she could not comprehend such an act and asked me to show her what I meant. I took my hat off and she started sobbing. Then she slapped my face and scratched her fingernails down my neck and screamed for our mother to come. Of course my mother beat me. Why? Because both of them were terrified that my short hair might jeopardize the marriage. I kept trying to persuade them that my hair would grow, that this was not the end of anything. They produced a turban for me to wear and forbade me to take it off.

But the story about my short hair did not end with the planning of the wedding festivities. It was like I was under house arrest—no school, not even a walk in our garden; I had to stay in my room so that no one would see me, especially not my brothers, who would have lost their minds had they known what I'd done. I was expected to stay hidden away until my hair grew back. Desperate to get out of the house and back to school, I convinced my mother that my reputation at school was already damaged because of the fights I had been in, so having short hair would not likely make things any worse. She came up with a plan—a genius plan, I thought. As long as I said that there'd been an accident—that my hair had been burnt and had to be cut—I could return to school. A very small price to pay. So off I went.

During the singing of the national anthem at school the next morning, the principal spotted me and my very short hair and called me to come to her office. She wanted an explanation. I told her about my hair being burnt. She didn't buy my excuse. "Do you know that cutting your hair is a violation which gives me the right to dismiss you from school and that will ruin your academic record, which will mean that no school will accept you and you will be forced to sit at home without any certificates?" I wasn't surprised by the diatribe, but it did reinforce my thinking about the value of girls in this place. I told the principal that it was not fair to destroy the future of a girl because her hair was short. She snapped right back with, "It is not normal for a girl to act like a boy."

I suppose I could have let it go and taken her threat as my punishment, but I invariably strike back at injustice, and this time I shot back, "Are you saying that I act like a boy because of a haircut?" She told me to get out of her office and threatened to fine me if my hair didn't grow back to a suitable length. On my way out the door I said, "This is the stupidest thing I have ever heard. What if my hair does not grow? It's not my fault." Then she said she'd call my mother. Again, I wasn't going to leave her with the last word, so as I closed the door I said, "Don't you dare bring my mother into this because of these ridiculous rules."

It didn't end there. When I told my classmates what had transpired, their response was, "Rules are rules and they're not meant to be broken." My hair had become a crime. Everyone in the school was talking about my hair. This was the beginning of a downward spiral that eventually swirled me into a depression. I began to see the people around me as curses in my life. I realized I couldn't live like this and told my girlfriend about my secret desire to go away, to live abroad. She seemed a bit shocked at first and asked if I was serious. But when I explained that I wanted to escape all the things that are forbidden, she liked the idea and reminded me that even our relationship was haram, and we could be killed for it. That's when we began to dream together of escape. I cut off my friendships with the other girls at school and spent all my time with my girlfriend as we plotted a plan to get away. But all the time I knew that if I waited too long, my family would marry me off to a man and I would not be able to run away with this girl who I now realized I had fallen in love with.

As if on cue, in the midst of my own conundrum, my father came one evening to tell us that he was going to marry again. The whole family was at home. He told our mother first and then called all of us to the living room to share the news. He sounded triumphant when he said, "I'm marrying a third wife." My poor devastated mother tried not to look at him or at us. Mutlaq and Majed thrust themselves into his arms to congratulate him, as though he had just scored an impressive win. Fahad was silent. That's what I loved about my little brother; he was an observer and could maintain silence while absorbing the consequences of injustice. Lamia rushed immediately to our mom and tried to console her while she wept and purposely kept her back to the room. It was a repulsive sight—the wounded woman warrior and the conquering hero man—the consequences of a mean and lopsided system. My father left with his two older sons by his side; my mother asked us to leave her alone.

The girl my father married was in her late twenties, maybe early thirties—young enough to have been an older sister in our house. Although men having more than one wife is common in Saudi Arabia, in parts of the country marriage is becoming increasingly monogamous as incomes decline and Western ideas about mutual compatibility between a husband and wife are taking hold. But in Ha'il, where I lived, that was not the case. The marriages are all arranged and usually cousins marry cousins. If a woman is divorced, she cannot marry again. Men can remarry as often as they like.

Although my parents were not divorced, my mother didn't want my father in the house again after he married his third wife. She only partially won that argument. My father didn't sleep at our house after that, but he claimed that the part of the house that was for men only—the sitting room—was still his private territory, and he invited his friends to gather there whenever he chose to. The main part of the house was separated from the sitting room by a door that could be closed and locked, so it seemed that he was in another house, but we all knew his life was going on as before while our mother was on this side of the door weeping. When he came to visit us, he had to come via the front door of the house. It was complicated and difficult and usually unpleasant.

While my mother was devastated and her confidence took another hit, my father's marriage affected me personally and made me question the men in my family and what their values were. Of the dozens of fathers and brothers in our clan I could only think of two—my mother's brothers—who I would consider decent men. My Nourah Mom had raised good men. One of them was always like a dad to me, treating me as if I were his own little daughter. I recall once when I was young and sick, he came and carried me everywhere on his shoulders. It was wintertime, unusually cold outside, and he put his big warm coat on me and tried to convince me he wasn't feeling the cold, even though I could see his hands were shaking. When I was older and wanted to go shopping, my brother would often say he didn't feel like taking me, but my uncle would come without hesitation and he would protect me like I was something precious to him, taking my arm to cross the street and always smiling and encouraging me and building my confidence. I had a wonderful relationship with my uncle; I don't know another man in my country like that.

By now my father had two more wives, one new baby with the second wife, and a first wife who was suffering from depression in a household where no one was sure about who was in charge. One day I came back from school with traces of kisses on my neck that I had forgotten to cover, and while I was talking with my older sister she noticed the traces, came close to be sure, and suddenly started yelling for my mother to come, as though there was an emergency. Mom raced to the room, and when she saw what the so-called crisis was, she showed no reaction to me at all and simply said to my sister, "This is not her first time and I don't know how to control her." And she left the room. Lamia must have thought it was now her job to torment me, and she started scolding me and calling me a whore.

After that, Lamia would drop into my room randomly to check on what I was doing and call me names to let me know she didn't approve of me. She told the driver he had to pick me up at school because my brother was often late and she didn't want me having any opportunity to hang around with the other girls. I felt like the walls were closing in on me. There was no safe place I could be myself.

In March, a friend of mine gave me a gift for my birthday. We don't celebrate birthdays, and the giving of gifts is taboo in our culture, but the guitar she bought for me made me very happy. Of course it was haram, so I decided to hide it in my room and teach myself to play when no one was around. And of course I got caught—everyone in the house was rummaging through my things looking for reasons to punish me. This time it was Majed. He was waiting for me when I came in from school and walked with me to my room. My guitar was lying on the floor; Majed pointed at it and demanded to know where I'd gotten it. Then he pulled me to the side of the room and bashed my head against the wall, demanding I tell him the name of the boy who gave me the guitar. He wouldn't stop beating me until I gave up the name of this phantom boy. Joud came into the room and, trying to save me, said, "I saw her friend come and give this to her. It was not a boy." He seemed to calm down but insisted I give him the name of the girl who gave me the guitar. I told him her name. With that, he picked up the guitar and smashed it down on my head. It was such a painful blow that I wondered what had broken—my head or the guitar or both. I wanted to remind him that he'd found the guitar because he was invading my personal space and that he should be ashamed of himself, but I decided my head hurt enough. I'd keep my thoughts to myself.

In the meantime, my girlfriend and I continued to plan for our escape from Saudi Arabia. We didn't know which country would suit us, so we started looking at the places I'd read about. I also discovered the word *refugee* in my readings and found out what that meant and how it might apply to us. There was a lot to learn—travel arrangements, obtaining a visa, figuring out refugee claims. I'd started gathering information and planning this huge exodus when somehow my body began to stop, slow down, feel like lead. The pressure I was feeling from my family, the school, our society in general began to get to me in a way I wasn't able to handle. Even though I was planning an escape, I started to feel overwhelmed by being under observation all the time, by the threats that came from my mother and sister, and by the peculiar things going on in my home—such as Reem's mental condition and Majed's domineering and hurtful way of treating our mother and me. I wasn't able to be myself, and I started to feel as though I didn't want to carry on living like this. I stayed in my room, lying down the whole time, looking at the ceiling. Only Sasha the cat was with me. She'd rub her head on my face, as if asking me to get up. She seemed to know something was wrong with me; she even licked my tears when I cried. But most of the time she lay on my chest and watched me.

I was so depressed I didn't know what to do. Every day was worse than the day before. Then Mutlaq came into my room. I didn't even have the energy to be afraid of him. When he asked why I was staying in bed, I told him that I needed help, that I needed an appointment with a psychologist. He was hesitant because in our society getting help like that is seen as a weakness and not acceptable. Furthermore, there aren't many clinics in Ha'il, and the available ones are a long distance away. But he knew I was in trouble and said he would make arrangements for me to speak to a therapist over the phone. I agreed. He didn't want my parents or siblings to know about this and paid for the sessions with the psychologist himself. It was unusual for him to help me this way, and I wasn't sure I understood his motive, but I needed help and accepted it without question. The therapist told me about drugs that would help and didn't require a prescription, so my brother bought them for me and I took them for several months without anyone in the family knowing.

❖ ❖ ❖

Lamia's wedding was approaching, and the family and relatives were absorbed with the planning and preparations. Lamia was a beautiful bride in a dazzling white gown and gold and silver jewellery. There were more than four hundred guests invited to the ceremony that began at 8 p.m. and went on until 4 a.m. The hall was beautifully decorated and the food tables were spilling over with meats and sweets. It was lavish to the max. But as required, the men and women were in separate rooms—we ate and danced separately.

The bride price the groom paid for my sister was 150,000 Saudi riyals (US$40,000). The money is a religious right called mahar, and its value varies depending on the financial ability of the man and the status of the woman. If she is a widow or divorced, she is less valuable. I saw the joy on Lamia's face and asked her if she was really happy with the price given for her. She said, "Of course I am, there's nothing wrong in this, it's my right."

I didn't say anything to upset her but thought how insulting mahar is—a strange man who she doesn't know pays an amount of money calculated by my father and buys her, as though she's a vehicle or a house. I had read comments on social media about mahar that said a bride becomes the property of the groom and he then has the right to do anything to her, even rape her, because he bought her.

I had to wear a wig to the wedding party so that no one would know that my hair was still short. My sister, dressed in her expensive and gorgeous gown, had to stay in a room by herself, which is our tradition. If she showed herself or her joy she would risk being called a shameless woman. In some cities the bride and groom come together at the end. But not in Ha'il. When the last guests left, my brothers walked

Lamia to her husband and the new couple left for a hotel.

My life went back to staring at the ceiling in my bedroom with Sasha beside me playing with my fingers as if to say, *Come on, get going.* The medications I'd been taking had run out. I was so sad and despondent and I'd lost a lot of weight, which normally would have been something that made me happy, but being so lethargic and pale, I hardly noticed the changes in my body. I wondered if I would ever be okay again. I didn't want to leave my room, but I knew I needed more medication, so I used my phone to call my mother and ask her to come right away. She did, and when I gave her the empty medicine package and asked her to refill it, she presumed I'd been taking illegal drugs and said, "Are you on drugs now?" I said, "No, it's not what you think. It's treatment for depression and my brother knows about it." The look on her face changed from disgust to amazement and she said, "If you had told me you were a drug addict, it would have been easier than knowing you're crazy and will be rejected by the world for your illness." She left my room and told everyone in the house that I was insane, and that now they would never find a man to marry me.

A few days later my dad came to my room. He found me with a turban on my head to hide my short hair, which was the least flattering thing I could imagine, and looking emaciated and feeling dejected. He sat on the side of my bed, took my hand in his and said he knew that I was suffering from depression. He wondered if someone had done this to me, if I was attached to someone who had hurt my feelings. He was trying to get me to open up about my emotional life; I wasn't about to do that, but I did confess that I was simply incapable of joy and all I wanted was to leave this life. He started to cry and, with tears pouring down his face, this man who had rejected all of us for not one but two more wives wrapped his arms around me and kept assuring me that everything would be okay. He said he would bring me vitamins to cure my depression and that I should leave those other drugs alone. I agreed to do that for him, but I knew my depression was getting worse and there was no one in the house to support me.

One day when I was feeling particularly low, I saw my mother sitting at the door of my room, chatting to a friend on the phone and keeping an eye on me. I picked up a knife and positioned the blade over my wrist and said loudly, "I will end my life now." She put the phone down, looked squarely at me and said, "You're crazy. Finish your life. You will go to hell for killing yourself. Allah will not be pleased with you, so how can I be pleased or satisfied with you?" As I cut into my wrist, the blood came spurting out, spilling everywhere. I felt dizzy. All my mother did was call out to my little sister, saying, "Go give your crazy sister some bandages." Joud brought me towels, alcohol and bandages and sat with me until I calmed down. It was the last time I tried to get my mother's attention.

Eventually the depression began to lift. When it did, and I recovered, I went back to my old habits, sneaking out of the house and getting together with friends—both boys and girls—that I met on social media. This was a different crowd. My friends at school would never go out like this and didn't know any of these people. But even at the beginning of meeting these rebels like me, I was careful about the ideas I shared with them; the parties they held took place in secret, and although I attended them, I thought they were scary, too wrong for me. It was a secret world in Saudi Arabia, a world occupied by minor girls and older men who were distributing drugs and cannabis to everyone for free. The government knew about these parties, so it would have been easy for them to raid the party and arrest us, but they didn't. I saw people at that party who worked with the religious police and other famous people from social media who were known for the advice they gave about honour and religion—all of them secretly fooling around with minor girls. I figured this was beyond what I was willing to risk and decided to stop attending the parties. This wasn't about breaking the rules that I didn't approve of; this was about flouting the rules and indulging in different but equally despicable behaviour. It was about hypocrisy and, in my opinion, it contributed to the duplicity I despised. So I quit that crowd, but it was an experience I will never forget.

There's a story we all knew about but were careful not to comment on—a story that illustrates the enormity of the double standard in our country. I was only two years old when a fire broke out in a girls' boarding school in Mecca, but the consequences of that catastrophe cling to our lives like a parable. It began on March 11, 2002. The blaze was reported to have started on the top floor of Makkah Intermediate School No. 31 at about eight in the morning. Firefighters said it was caused by "an unattended cigarette." There were eight hundred girls registered at the school, most of them from Saudi Arabia but also international students from Egypt, Chad, Guinea, Niger and Nigeria. The school was overcrowded, and it didn't have the required safety features and equipment such as emergency exits, fire extinguishers and alarms. The flames spread quickly and the school filled with smoke. The girls, most still in their rooms getting ready for breakfast and morning classes, raced to the exits, but the guards posted there refused to unlock the gates so they could escape. Why? In their haste to leave the burning building, the girls had not dressed properly—they were not wearing head scarves—and their male relatives were not there to receive them on the street. The girls were screaming to be let out, and passersby stopped to help as the school turned into an inferno.

Then the mutaween turned up and beat back the crowd, reminding them that the girls would be committing a sin if they came out of the school without covering their heads. According to eyewitness accounts, they told the incredulous crowd, which now included parents of some of the girls, that they (the religious police) did not want physical contact to take place between the girls and the firefighters for fear of sexual enticement. Some reporters claimed that the few girls who got out were pushed back into the burning building by the guards and the mutaween. When the firefighters rushed inside to rescue the girls, they were also admonished by the religious police.

Fourteen girls died in the fire. More than fifty were injured. I shudder every time I recall that terrible story and ask myself what sort of barbarism was at work when innocent girls were forbidden to escape from a burning building.

This is all done in the name of honour. It's the silencer in every conversation. The phrase "we are clean people" is as common as saying good morning. Saudi society is fixated on honour—in whatever convoluted description works to cover the unequal and unacceptable behaviour of the people who own the power. To think that fourteen girls died a horrible death and fifty suffered such awful injuries—burns and broken bones—because of so-called honour is repulsive to me. But it means everything to my family and to all families in the kingdom. Saudi Arabians will kill—make that murder—for honour.

Those tensions seeped into our bones as girls growing up in this country. By the time I was sixteen years old, I knew the path I was on—to alter the course of my life and embrace my identity as a feminist and an atheist who opposed the laws of a state that deprived its citizens the rights of expression and freedom of lifestyle—was the correct one. My dream was to live in a country far from the Middle East that believed in gender equality and human rights. But I couldn't even convince my family to let me go to high school in another city in Saudi Arabia. I remember the day I asked my mother and brother if I could go to high school in Riyadh. My brother laughed at me, saying, "Are you serious?" while my mom interrupted the discussion and ended the conversation by saying, "We do not allow our girls to study away from their family." I did not argue with them because I knew that they would never consider my point of view. It was left to me to seek change.

CHAPTER FIVE

Secret Codes

During my last year of high school, I began to move away—emotionally—from the people, the customs and the laws that had confused and then infuriated me as I came to terms with the size of the penalty that's exacted for being a girl instead of a boy, that denied me fair opportunities and even put my life at risk. It took time for this separation to occur. Like a ship that is docked in a harbour and fastened by gangplanks to the land, I was attached to my family, to my country and even to many of our customs, like the way we stay close to our cousins and uncles and aunts. But just as a ship pulls up anchor and begins to slowly but surely shift off the shore, I too was cutting the ties that bound me to pretty much all that I knew.

That's a lot of poignant baggage for a seventeen-year-old to carry around. Not only that, but as I examined the conditions in my life that I now determined to be intolerable, I had to keep all of it to myself because telling would mean risking the wrath not just of my family but even the government of Saudi Arabia. Their response to my decision to reject what they stood for could run the gamut from shaming and shunning to locking me up or even killing me. Criticizing the religion, the government and the farcical centuries-old customs is not acceptable in Saudi Arabia. So although it was painful to hide what I was feeling for fear of this retribution, my own silence was the protection I needed. But it didn't stop my curiosity, which is what drove me back to the internet and the illegal sites that provided answers to the questions I had been mulling over for most of my life.

I was still searching online for information about people who were in a similar position: atheists, critics, feminists, homosexuals and people who opposed the government. They were perplexed, disgruntled, angry and, like me, they had tried and failed to change the laws that allow families to kill daughters, police to threaten and harass anyone they choose, and the government to refuse to acknowledge human rights. As much as meeting people like this online was liberating and refreshing, this was also a time of psychological pain. I was coming to terms with a life of being forced to do what everyone else demanded that I do, of being beaten and oppressed by people who were supposed to love me, of being educated by teachers who thought they owned my body and my mind.

That was the state I was in when I stumbled upon an online connection that led me to an extraordinary underground network of Saudi women runaways. It was bedtime. I was surfing around various internet sites and, by an amazing coincidence, found the Twitter account of a woman who lived in Canada. I texted her and asked her how she'd managed to get to Canada and what she did to get a visa. She replied right away and told me exactly what she had done. Her information was easy to follow. I texted her right back and, after a few more helpful messages, felt safe enough to tell her my story. I trusted her enough to admit that I wanted to run away. I shared my preliminary plan with her and said I was worried about whether or not it would work. By now I was sitting on the edge of my bed, fixated on my phone as though I was holding not, it but the hand of my rescuer. I couldn't believe I was having this conversation with a woman halfway around the world in Canada, a woman who seemed to know every-thing I needed to know about escaping. She asked me a few questions to make sure I was determined to leave. Then she told me that there was a group on a social media site who had the same feelings and well-thought-out ideas and who knew how to help each other. She explained that they were all girls, most of whom used fictitious names, and that they never shared personal information.

She gave me the secret code to access this site that helped girls escape and told me to create a code name for myself. I felt as though I had just received the keys to my own king-dom when I entered my code name—Sasha, after the cat that I adored—and first met the people in that undercover site who opened the world to me and became my lifeline. Without them I would never have made it to Canada. Once on this clandestine site, I was hooked, and I spent every free hour I had learning, examining, testing and, finally, plotting my escape.

This was a network of young women—a private chat room—who were like-minded in their views about the status of women and girls in places like Saudi Arabia. The girls I met there all shared the same issues and the same desire to either change the country we lived in or leave it. There were subgroups, some of them with men as well as women who were also interested in getting away and living in a foreign country. I felt safe with the group of girls. Even though our identities were hidden, they very quickly became a family to me. We shared information, planned together, worried about each other. When one girl recounted a story about her father beating her and denying her one human right after another, the shared angst for her was palpable. Each of us jumped into the conversation with advice about the need to stay hopeful and how to go about making a change in her life. There were so many good ideas being shared, and so much support, that I felt a powerful sense of belonging. For example, they all had tips about how to get a travel permit without asking your guardian, who would never approve of such a thing. These women knew how to get themselves off Absher—the app that allows guardians to control the movements of the women in their lives—and get a travel permit secretly. It was a tricky procedure, and dangerous too, because the government would hunt down any Saudi national who tried to flee the country. The girls provided excellent tips on how to trick the government to get papers, how to know when to make a run for it, and when to use a fictitious name and number to cover your tracks as you created your escape plan.

For me, it was like a homecoming: I was no longer the odd one in the group, with ideas that challenged the status quo. The people I met on this chat site were mostly from Saudi Arabia and the United Arab Emirates, but also from other countries that use the male guardian law to control the lives of women and girls. Some were ex-Muslims; others had already escaped. But everyone on the network was bound by solidarity and secrecy. They were all against the guardianship law, and although some planned to stay in their countries and fight for change, others were looking for an exit from the state that mandated a crushing repression in their lives. Most didn't know each other except through the online meetings, but the bond they had built with one another was immediately obvious and the trust factor powerful and useful. It was a support group that worked to help everyone who joined, at whatever stage they were at in their lives. With nowhere else to turn for advice and information, they gathered like supersleuths online and shared their knowledge and tips and support, one to the other. They sustained me and made me feel my dream was possible, that all was not lost.

The stories they told were as inspiring as they were encouraging. One said she had to leave the day before writing her final exam in her last year of university because her escape hatch was closing. Her friend, who was also planning to run away, was arrested. She feared staying even one more day, which meant giving up the degree she'd studied for during the last four years. Another described numerous carefully planned escapes that she had to abandon at the last minute because of an unforeseen glitch. And one girl who escaped still had braces on her teeth; she had to have them removed once she'd claimed asylum. The ones who had already escaped told vivid stories about the snags in their plans and the perils they ran into. That's how I found out that the border guards in Australia might ask to speak to my father when I arrived. The network girl who had escaped to Australia told me to make arrangements with a boy who would take the call and speak as my father. These networkers had planned and presided over many escapes and had all kinds of useful tips—such as buying some time in the initial stage of the escape by removing the SIM card in my phone and relying on Wi-Fi in airports and cafés so no one could trace my whereabouts.

Until meeting the others, I thought Saudi Arabia was the only country that treated its women like this. In fact, I always saw places like the United Arab Emirates and Kuwait as being much more progressive countries, where the rights of women had made significant strides forward. But ultra-conservative families, the abuse of women and girls, and governments that hand the control of women over to male guardians are common in these places as well.

There was a lot to learn: how to book an airline ticket online and apply for a visa without anyone knowing; how much money to save and how to stash it away in a secret bank account so no one would know about it. Escaping is the toughest step but not the only one to consider. Once arriving safely in another country, you need to know how to apply for asylum, and how to continue your education or find a job so you can pay your way when the money you have saved runs out—these are all critically important steps. The network has that information and feeds it out online like training courses—first this, then that. It taught me almost everything I needed to know. But apart from being a digital how-to-escape guide, this is the organization that pumped up my confidence in myself and assured me that my decision to abandon all that I knew was the right one. And the women all related to the perils involved in eluding family and authorities when secretly booking travel and carefully deciding when to run and where to run to. Authorities in Saudi Arabia are naturally suspicious and are always on the lookout for girls who step out of the lines they were raised to stay within. These women knew that so-called honour was the noose around all our necks.

As time went by, I became involved in the escapes of several girls. I think it's fair to say that together we shared the exhaustion and the psychological pressure whenever an exit plan failed. But the group was steadfast in helping each girl every step of the way—whether stealing your father's phone to get access to your personal information or finding supporters in the asylum country. Even if you're not a well-known Saudi, the embassy of Saudi Arabia in the country you choose will contact you, try to lure you back and eventually try to have you arrested. Three of the women in the group I joined became my own tightly knit support unit when I came closer to acting on my plan. They were the ones I was texting from the car in Kuwait and from the airport before I left for Bangkok. We had become each other's best friends.

Some of the data I gathered came from a website. The site is visited mostly by women like the networkers I ended up with, those between the ages of eighteen and thirty who are desperate to leave the abuse in their lives and begin anew in a country that's safe for women and girls. Most of the information on this site is from asylum seekers; it's about how to talk to authorities, ways to get your hands on your own identification papers (such as your passport) and how to navigate each step of the escape. The authors understand how frightening the journey is for eighteen-year-olds who have never travelled on their own, never spoken to a man outside the family—never mind a customs officer with complicated questions. There's a lot to deal with for a girl like me, who's been driven to school all her life, who's not only been supervised while shopping but hasn't even been allowed to hand money to the shopkeeper. In the chat room, the networkers try to teach you and even practise with you so you know how to handle yourself when meeting strangers or if—heaven forbid—you get caught and taken to a detention centre. They remind you that just because you see one country or another as a safe haven, you need to watch out for criminals everywhere. This network, like the clandestine site I'd been on, was another extension of this family of activists—the people who care about your safety, your future, your happiness.

If I could name a single issue that drives women away from our countries and binds us together under the mutual banner of the tormented, it would be the male guardianship law. It is the bedrock for all the discrimination and anti-woman violations and human rights abuses that follow. I discovered in these online chats that although Saudi women have always been less visible than Saudi men, these draconian guardianship laws were not always in effect. Until 1977, women were allowed to travel without a male guardian. But then a princess called Mishaal bint Fahd made the mistake of falling in love. She was the daughter of Prince Fahd bin Muhammad bin Abdulaziz Al Saud and a granddaughter of Prince Muhammad bin Abdulaziz, who was an older brother of King Khalid. For all her connections to princes and kings, the government killed her for her crime of loving a man of her choosing. And then they made a law to make sure no other girl would ever be unaccompanied by a guardian and therefore able to commit the crime of falling in love.

The princess had been in Lebanon attending school.

During a courtesy call to Ali Hassan al-Shaer, the Saudi ambassador to Lebanon, she met Khaled al-Sha'er Mulhallal, the ambassador's nephew. The story starts like any presumably normal girl-meets-boy story. But this is a Saudi story, so there's nothing normal about it. They dated each other, met secretly and eventually had an affair. They were safe as long as they were in Lebanon because of that tried-and-true shibboleth: if nobody knows, nothing is done. A family's honour is tied to public "knowing," and in Lebanon, no one knew. But once she was back in Saudi Arabia the scandalmongers churned out the story and then shame became a wound, and the death of nineteen-year-old Mishaal was the salve used to soothe the family and stop the gossip. That's how it works. She tried to escape. She even faked her death, making it look as though she had drowned, and then dressed herself as a man; she made it as far as the airport before an official at passport control saw through her disguise and sounded the alarm. She was returned forthwith to her family. They wanted blood. And of course they extracted it with the usual Saudi duplicity. They used the Sharia legal system to murder their daughter. It didn't matter that Sharia law requires the testimony of four adult males who claim they witnessed the actual sex act, or a confession of guilt by the victim (who must proclaim "I have committed adultery" three times in court). The reports say her family told her not to confess but to vow she would never see her boyfriend again. They claim she stood up in the court and said, "I have committed adultery. I have committed adultery. I have committed adultery." Of course, there is no record of a trial or her confession.

Both Mishaal and the man she loved were executed in an act of revenge for the dishonour she brought to her tribe: she was shot; he was beheaded. But the story doesn't end there. Like everything else in my country there are two versions—and no source for truth. The version we talked about in our online chats came to light three years later, when a film crew investigated the love affair and made a movie called *Death of a Princess*. The filmmakers claimed that even the execution of the pair was done without the process of tribal law or religious law, but what followed really told of the lengths to which Saudi Arabia will go to hush up its murderous acts. Apparently, the movie was to be broadcast in Britain on ITN and in the United States on PBS. But a virulent protest threatening diplomatic, political and economic penalties if the film was not cancelled hit the producers. It's been said King Khalid offered a US$11 million bribe to stop the airing of the film. It did air, but there were repercussions. The British ambassador to Saudi Arabia was expelled, and the Mobil oil corporation—a major sponsor of PBS programming—took out a full-page advertisement in the *New York Times*, claiming *Death of a Princess* would damage US relationships with Saudi Arabia. The ad read, in part, "In Saudi Arabia's view, the film misrepresented its social, religious and judicial systems and, in effect, was insulting to an entire people and the heritage of Islam." And it read, "We hope that the management of the Public Broadcasting Service will review its decision to run this film and exercise responsible judgment in the light of what is in the best interest of the United States."

In the chat room, we also talked about the many Saudi women who were hidden away, imprisoned, or killed by their guardians, and about some who committed suicide. The names of Hanan al-Shehri, Khadijah al-Dhafiri, Amna Al Juaid and, of course, Dina Ali, stay with us. Hanan al-Shehri, twenty-five, committed suicide, but her sister Aisha claims that a family member had been beating Hanan and threatening her the day she died. She had called the police, but they insisted she get medical documents to prove she was being abused by this family member. Then she died in a fire set in the backyard. Her sister also says the fire that took Hanan's life was not set by Hanan. No one who knows her believes she committed suicide. But the police still say Hanan killed herself. We don't believe them.

Amna Al Juaid is another Saudi girl who escaped from her father's house two years ago after claiming her father beat her and threatened to prevent her from going to university because she refused to marry her cousin. She went to live with a foreign family, got a job and began living independently. Then her father found out where she was and began negotiating with her to come home. He said he'd give her passport to her if she came back. She recorded her own story on a video and asked her friends to post it if she disappeared. It's been posted. Human Rights Watch urged Saudi authorities to investigate. No one has heard from her.

Khadijah al-Dhafiri died at the age of twenty. She'd been horribly abused by her husband and eventually jumped off a third-floor balcony to escape him. She survived the fall, but her severe injuries led to paralysis and eventually she died in the hospital after suffering a heart attack. Basically, her husband is responsible for what happened to her. However, he served only four days in prison and walked away because she was the one who jumped.

These are some of the stories of the many women who suffered at the hands of their guardians in a country that executes activists and frees criminals.

❖ ❖ ❖

I was absorbing data from this group like a sponge. All of it fed into what I had suspected, reinforcing my belief that there had to be change in this country. And it underlined my dilemma: I was an outsider in my country, an infidel to my customs and an undesirable to my family.

In the meantime, my life at home and at school carried on as though nothing unusual was occupying my mind every waking hour. My brothers stayed in private schools while I went to the public high school. Their lives were so different from mine, it was as though we lived on two separate planets. It wasn't just the freedom they had to be outside with their friends, or the right they had to swat me across the face just to show me who's boss, or the liberty they took with our mother, telling her to be quiet, to calm down, to stay in the house once she was vulnerable; it was all of that and a lot more. My brothers could travel. They could go out in the wide world and see the sights of a foreign place, taste its flavours, find out who lived there and what their thoughts were. How I longed for such opportunities! I was hungry for knowledge. After I'd been on the illegal websites and had seen how people lived their lives and discussed forbidden topics, and as I watched their actions in support of their beliefs, all of this astounded me and stirred my soul. I wanted to be part of the world. I wanted to be out there asking my questions, sharing views with others and having fun instead of being forever silenced and subdued and told girls "don't do that."

My brothers were driving their own cars; they'd actually started driving when they were thirteen years old. Mutlaq was attending university—a very religious one—in another city. But at home nothing changed. By the time I was seventeen I had very little interaction with them, but I was very close to my younger siblings, Fahad and Joud. Although it wasn't quite as simple as that: Fahad had morphed into a son who controls the women in the family as soon as he was a teenager, but he exercised those rights on Joud and not me. He and I had always been best friends, maybe because he was so sick as a child and couldn't go outside to play with the others. He'd had to stay in the house with me, and we grew up together watching television, playing our own games, drawing and comparing the artwork we had done with each other—basically, we were being nice to each other, which was a fairly unusual trait in our house. Playing together ended when we got older, because he would have been ridiculed for spending all his time with a girl, but I could always count on Fahad to help me get out of the house. He'd come with me as my guardian and we'd go out to eat in restaurants or go shopping at the souk or just sit in the park for hours and talk. He was two years younger than me but had a car, so we could take off and go wherever we wanted—playing music in the car, laughing out loud and doing what kids our age are supposed to do. I loved being with him, and I always wondered about the difference in the relationship I had with him compared to the one I had with my older brothers. I knew in my heart that siblings should care about each other, should stick up for each other, make special memories together. I had that with Fahad. But never ever with Mutlaq or Majed.

Oddly enough, or perhaps it's a telling detail, my mother disliked the relationship I had with Fahad. She insinuated on more than one occasion that he and I might be sexually involved with each other. She would suddenly appear in a room where we were talking or push open a door as if to surprise us when we were watching television. For me, her actions just reinforced the notion that girls are meant to be distrusted, that girls being happy and sharing their affection for their siblings was not the norm. Eventually—as though my mother was reading a young boy's mind—Fahad did start touching me inappropriately. The first time it happened, I froze. I didn't know what to do and sat there dumbfounded, staring at him and wondering what would happen next. He saw the look on my face, and I assumed he understood that I was horrified that my sweet, gentle brother was taking advantage of me—only because he was a boy and I was a girl. A few days later when we were alone, he did it again, this time not just touching my breasts but other parts of my body as well. So I told my mother. Her words were plain: stay away from him and dress more conservatively. She asked, "Did he only touch you or did he do something else?" These were my mother's parameters when it came to sexual abuse—Fahad could touch me without my permission but could not have sex with me without her permission.

As much as I was truly astonished by Fahad's advances, I wasn't surprised by my mother's interpretation of what had happened. She blamed me! She said I must have provoked him, must have wanted him, acted like a whore and misled him. Then she turned into an inspector and started watching me. It was more like spying, actually, and not just on me but also on my little sister, Joud, so that there would be no evidence of Fahad touching either of us. I believe this was about protecting Fahad's reputation and not mine. These incidents, if made public, would bring shame upon Fahad and, in turn, our family—not because of his bad behaviour in touching me, but because of whatever I had supposedly done to tempt him. Thankfully, Fahad got the message and backed off. Our close friendship went back to what it had been, but the experience stayed with me like a warning.

Although I was happy to have my baby brother back, I certainly never shared with him my biggest secret—that every night when the family went to sleep, I was entering a concealed code into my phone and meeting with my new undercover family. I guarded my secret like a satchel of precious jewels, storing my newfound knowledge away, gathering what I needed to know and biding my time.

By now Lamia was married and had moved away from home. But Reem was still there and going to the same school I was going to. Ever since the incident in her bedroom when she had taken my father's gun and tried to run away, she hadn't been the same. She'd been a terrific student before that but now had difficulty at school. She'd been outgoing and popular; now she was timid and withdrawn and needed the rest of the family to take care of her.

I knew from what I was reading online that families in other countries didn't live like this, that daughters weren't treated like shameful additions, that girls had the right to live their lives as the boys did. As a teenager in Saudi Arabia, I was supposed to spend all my time at home. Although I did sneak out, I wasn't allowed to go out or to have fun. There were no activities for girls; we didn't even know what hobbies we might like because we never tried anything. Every day, all the time, I heard over and over again from my family, my teachers, even my friends—home is the right place for girls.

Life for a teenage girl in Saudi is determined by a series of strict rules: I cannot leave the house I was raised in until I marry and move into the house of another man, who would then control me as my father and brothers had at home. As a young woman I couldn't even open a bank account without my guardian's permission. When shopping, I wasn't allowed to try on clothes, even in a fitting room. What part of this is sinful? I ask. And the advertisements in my country blur the faces of women but not the men. It's like a constant message. Repeat after me: *You are woman. You are invisible. You have no value.* I had to ask permission to go to the doctor. I knew students who wanted to be medical doctors or open their own businesses, but still had to take orders from their younger brothers. There are stories about women studying medicine in the United States and still being bossed about by illiterate male guardians at home.

I was living in a society that forbade me to speak my mind and saw my ideas for changing the rules as a criminal act. It was a form of being in prison, because your real truth is locked inside you. I didn't hate my family—I certainly never saw them as my enemies—but they were the architects of my future, which was set to be a dreary life that didn't allow me to fulfill my dreams and desires. I felt a home should be a place to feel safe and to have the right to speak and debate and express emotions. I wasn't going to find that in this house; by now, I knew that staying there was not the future for me.

There's plenty of evidence throughout the country that speaking for change will make your life a misery. Raif Badawi is a writer, blogger and activist who made headlines when he posted an online message that said, "To me, liberalism means simply to live and let live." It's dangerous to speak about liberalism in Saudi Arabia, but it didn't stop him. He also highlighted the plight of women, and questioned why women needed a male guardian to walk down the street and why it was extremely difficult for women to access the labour market and employment.

Raif was arrested in 2012 for "insulting Islam"—a life-threatening charge. He was eventually sentenced to seven years in prison and six hundred lashes. In 2014 the sentence was changed to ten years in prison and one thousand lashes, as well as a fine. After the first fifty lashes, the next flogging was postponed because of his health. His wife, Ensaf Haidar, who along with their two children took refuge in Canada, said her husband could not survive more lashings because he has hypertension. The world was watching this case. It still is. However, Raif Badawi is still in jail for speaking out about fairness and justice. The whole world knows, but which country is willing to cut ties with Saudi in order to call out these abuses?

Saudi women learn to subvert the stifling rules so they can survive and stay sane. Plenty have secret apartments they rent from agencies that waive the guardian rule, a place to go to feel free. Some girls take driving lessons from clandestine sources while waiting for the ban against women driving to be lifted. There are even secret soccer leagues for girls. I once asked a teacher to explain the ban on sports for women. My mother had suggested that being active in sports would destroy my virginity. I wondered if being inactive would destroy my health (obesity is a serious problem in Saudi Arabia). But I got the same old tedious reply: *You're a girl. Be quiet. Obey. Don't ask so many questions.* Some soccer stadiums have started allowing women in to watch the games—as long as they sit in a segregated corner of the stands called the "family quarter." None of that happened in Ha'il.

But there is evidence of change elsewhere in the country. The year I was seventeen, Princess Reema bint Bandar, who was the director of the Saudi Federation for Community Sports, said, "I've encouraged women to go out on the streets and into the public parks to exercise. I've been telling women they don't need permission to exercise in public, they don't need permission to activate their own sports programs. And more and more they are doing it." Maybe this was in response to an incident the year before, when Malak al-Shehri tweeted a picture of herself without a headscarf and was arrested. Not only that, but there was a public hue and cry for her to be executed. Whether or not reforms such as women playing soccer and driving cars or being out on the streets will stick is another troubling point, because most of the reforms are at the whim of the Crown Prince Mohammed bin Salman, better known as MBS. He presents as the reformer who granted women the right to drive, but he is also responsible for jailing the women activists who lobbied for that right.

The bottom line in Saudi is this: legally, a woman is a nullity. She can vote, but in court her testimony is overturned by a man's because in Sharia court the testimony of one man equals that of two women. Try fighting a case of abuse or assault in that sort of court. And if you aren't Muslim, forget it: in Saudi you have no rights in court at all.

There are examples throughout the short history of this ferocious place that a girl's life is seen either as a reproductive tool or a bargaining chip. For example, King Abdullah bin Abdulaziz—who had thirty wives and about thirty-five children before he died in 2015—performed his royal duties with aplomb but used women like disposable possessions. One of his wives, Princess Alanoud Al Fayez, was married off to the forty-eight-year-old king when she was just fifteen years old, without ever having met him. She gave birth to four daughters but no sons and her life became a misery. Kept in the palace as though she was part of a collection, she found herself in a long line of other wives to compete with and eventually decided her life would be better fulfilled elsewhere. She divorced the king and fled to the United Kingdom. To punish her for not giving him a son and for escaping to the West, the king put their four daughters under house arrest for over a decade and refused them permission to leave the country.

To be fair, King Abdulaziz is also known for his philanthropy; he's donated US$500 million to the World Food Programme and US$300 million to rebuild a New Orleans high school after Hurricane Katrina slammed into the United States. But his ability to alter the lives and livelihoods of his daughters because he's mad at their mother is a tale as enduring as the sands of Saudi Arabia.

My life at school continued as before, but during my grade twelve year—my final year in high school—I had a teacher who left a lasting impression on me. She saw me as different from the others. I'd heard that before. My mother kept telling me she wished I wasn't so different from my sisters, and my brothers insisted I had to stop being different and punished me harshly for not being like the other girls. Only my Nourah Mom ever said I was different in a way that made me feel good. Now this teacher was doing the same. I knew I was different, but even with the immense criticism I felt from my family, I always thought being different made me stand out as special, not as damaged. This teacher reinforced that thinking.

She saw my behaviour as proof that I needed more attention than the others. She told me she was trying to understand why I was breaking rules, and not afraid of anyone. She once told me that she felt I had excess energy at school because I couldn't be who I wanted to be at home. And that was true. I could never express myself the way I wanted to at home. But in her classroom I found a voice that I embraced. The course she taught was fun; she gave us work that asked us to describe ourselves and who we thought we were. It was the first time I could say what I was thinking without getting into trouble. She would ask questions that made us talk about our opinions. For example: Do you like wearing a uniform? When I answered and I expressed my feelings and thoughts about wearing uniforms—I hated wearing uniforms because we looked alike, and nobody could wear whatever they wanted—those conversations allowed each of us to show our personalities. That led to conversations I wanted to take part in, rather than boring lessons that I preferred to ignore. She looked at my work with interest and gave me good advice, and she would let me speak and share my answers in front of everyone. I'd marvel at the end of the class, thinking, *She's actually interested in me as a person, as someone with something to say.* I loved those classes.

Being happy at school opened other doors for me. It gave me the confidence to be a leader rather than a troublemaker. One time we were asked to create a project. The assignment was to describe the reason for choosing this particular project, the way to proceed one step at a time, and the results we each expected. I dove into this with gusto, choosing to create a charity to provide winter clothing for poor people. It was easy to describe the reasons for the project, and I knew precisely what I had to do: identify the people the charity would serve, collect the money to buy the items of clothing they needed, and deliver the items to each family. Then I could test my results by seeing them wearing warmer and more suitable clothing during the winter months and report back to the teacher about the number of families I had served. I was astonished that my classmates wanted to help. We started the charity together and they asked me to be the leader of the team. Everyone in our class wanted to make a donation to this cause, so there was plenty of money to buy what we needed. It was one of the rare times when I felt the flush of pride in myself.

Mind you, my education, especially during my last year of high school, was also peppered with indoctrination. I call it the "good wife" story. The teachers taught us incessantly that a good wife stays at home to cook and clean for her husband, and that Allah and the angels would curse us if we refused to have sex with our husbands. Lessons like this were more than I could abide, so invariably I would ask questions such as, "How can it be that Allah insists on forced sex? That's rape." The teacher said, "No, Rahaf. There's no rape between the husband and the wife!" Then, as if to reinforce her lesson, she kept telling me I should not refuse sex with my husband—first, because it's his right, and second, because he would cheat on me if I refused. "What about my rights?" I wanted to know. The girls in my class looked down when I asked these questions. I could not understand why. Were they shy or scared of what the teacher was talking about? Or did they look the other way because they agreed with her?

The teachers also taught us that a husband has the right to beat his wife as long as the beating doesn't harm her too much. They quoted from the Quran (4:34), which says, "But those [wives] from whom you fear arrogance—[first] advise them; [then if they persist,] forsake them in bed; and [finally,] strike them. But if they obey you [once more], seek no means against them."

Being a good girl, a good wife, was the mantra we learned year after year at school. The caveat was all about protecting our reputation so that we'd be selected by a prominent family as a bride. The teachers would say, "If you ruin your reputation by doing something haram, the man's family will find out when they come to the school to ask about your behaviour. If you have a ruined reputation no one will have you as a wife." I heard this threat ad nauseam throughout my three years in high school because most teachers thought I was a bad girl.

My family thought the same, because I never gave up on disputing the notion they had that girls were less valuable than boys. That meant my life at home still had all the strain that comes from unhappy relationships. My father continued to travel the nearly two hundred kilometres to Al Sulaimi each week and come back to Ha'il on weekends. All three of his wives lived in Ha'il. I used to wonder how he would decide which wife to stay with on his days back in Ha'il. Did they take turns? Did one wife offer something another wife denied? There must have been a pattern of sorts, or maybe just whims that were decided during the drive back to Ha'il. The process fascinated me, but I would never have dared to ask my father or anyone else for answers to my questions. Maybe there was a master plan, because he fit us in as well. Sometimes I saw him once a week, sometimes many days in a week and sometimes hardly at all. Again, I wasn't privy to his plans or his thinking about how to divide himself up among so many people.

By the time I was in my last year of high school, the second wife had a daughter and the third wife had a daughter and a son. My brothers loved them and showered them with attention and affection, visiting the children every day that they were home and taking them out for car rides or visits to the park or into town for ice cream. I watched all this with a mixture of amusement and disgust. Is this how it works? Out with the old, in with the new? Will they beat that little girl senseless, like they did me, when she is old enough to become an object of distrust and embarrassment? Will they take the new son under their wing and teach him the way of the world by showing him how to invade his sister's room, confiscate her things and punish her for having something as danger- ous to her health and her future as a guitar? How did they feel, I wondered, about their own mother, the woman who had given birth to them and allowed them to become masters of her home? Was it disgust I saw on their faces when they looked at our mother, or was that pity? Perhaps they were in transition to becoming husbands and fathers, and the scowl on their faces would in time be replaced with the look of com- placent satisfaction that my father wore. As time passed, I was increasingly convinced that there was something enormously wrong with the established status of men and women in my country. My feeling as a high school senior was that this was not sustainable. There were too many people in this country who felt revulsion and disgust for the status quo.

My mother certainly didn't like any part of the new arrangement that she was supposed to—as a good Saudi wife—see as an extended family, but she couldn't leave it alone. She harped about it constantly and was either yelling at my brothers for their betrayal in seeing the other families or begging them to shun the newcomers and show their loyalty to her. It was a never-ending drama. For me, it wasn't so much a matter of who was guilty as who was innocent. I thought the children were innocent: it wasn't their fault that my dad took their moms. I didn't see them often, but when I did I was drawn to hugging them and playing with them. They were little and beguiling, as children are. My sisters had absolutely nothing to do with them—not with the wives or the children. Their loyalty to our mother was unbreachable. I scored points with my dad and my brothers for the attention I paid to the new family, but this was not an act I wanted to share with my mother. I hoped to keep it from her. Of course she found out. In my house everyone had secrets, but no one kept secrets. We were never the kind of close family that protected each other's inner thoughts. My mother was hurt by my actions and told me so. I tried to explain that these were children who had nothing to do with our family quarrel. I wasn't even sure that I believed that myself, but I did feel guilty about fraternizing with the enemy camp, so to speak, especially when she accused me of being a sellout. Her pain was so real you could practically touch it. She'd been discarded. She cried a lot; she didn't want to talk and kept to herself most of the time.

I was already on a mission myself, and to be honest, I didn't have room for my mother's angst while I was leaving my religion and trying to find a plan for leaving the country. I look back now and feel I was being selfish—she was in such tough shape with no one to take her side except my two older sisters. The relatives—the aunts and uncles and cousins—had to support my father because that's what all of their families did as well. Being cast aside was not supposed to be an agony. It was part of the cycle of life for a good Islamic wife!

It's important to me that people understand the insidious effect the religion can have on an individual. It's one thing to reject it like I did; it's another to believe and practise the faith, and still another to pretend to be a believer like many do and just go along with the pretense. But it's quite something else to become a fanatic, and there are way too many examples of that in Saudi Arabia. My brother Mutlaq is one. Even my parents struggled with his blind devotion. I don't know what it is that takes hold of people and makes them think everything is a conspiracy, that everyone is anti-Islam. They depict Allah as controlling, hateful, devious and punishing. Devout Muslims don't believe that, but fanatics do. They look for people to blame for their troubles, and they see outsiders as apostates, which is the equivalent of a death threat for Saudis. Why can't people choose their faith? How is it that a government can tell you what to believe?

I still struggled mightily with the paradox that was my life, and the debilitating depression that had consumed me the year before continued to skirt the edges of my life. There is an Arabic expression: "*Mwlam 'an yaetaqid klu min hawlik 'anak nayim . . . biaistithna' wasadatk hi alwahidat alty taraa haqiqataka.*" In English it translates to "Painful that everyone around you thinks you are asleep . . . except your pillow is the only one who sees your truth." The words actually mean that what you show to people is the opposite of what is inside you. That was a constant in my life during my last year of high school.

Sasha, my sweet caramel-coloured cat, ran away that year. By then we had three other cats, called Leo, Kato and Lusi, so I adopted Leo to have someone close to me. I wasn't allowed to go anywhere, of course—being a girl, I was stuck in the house, and being a depressed girl meant even more isolation and shunning. So I started sneaking out to meet my friends in cafés again. Although I'd done it before, now I was even more audacious as I crept down hallways and slipped out the door and scurried along the pathways to the café. I often made myself giggle thinking about how good I was at vanishing without a trace, like some sort of girl detective. I have to admit the act of getting away with these escapades intrigued me. No one was watching, and being with my friends was a huge reward for me. We didn't have to worry about being caught at the coffee shop because the workers were all from India. They didn't know our language and presumably didn't care about our stifling rules. But I knew the consequences of being caught very well. Although my mother was basically ignoring me by now and my brothers were away at school and only bring-ing their hopelessly oppressive attitudes home on weekends, I knew that they would relish the idea of beating me, as if it enhanced their virility as Saudi men. I would wonder what sort of manhood required its new recruits to beat up girls.

During the last high school term before starting univer-sity, I decided to join a gym and get into shape. Majed reacted to that news as though I had planned a murder. He went into one of his rages and forbade me to go to the gym. A lot of men don't accept the idea of women being involved in sports, and many clerics have denied women the right to participate in activities as well. I knew which weapon I had to use to make this happen—my mother. She was the one who could convince my brother to let me join the gym. It worked. But there was a condition: my brother asked my mother to go with me and wait until I finished my exercises. Every day she sat on the floor at the club door, waiting for me to finish. At first I was excited and happy that she was there for me, as though my working out at the gym made her think better of me. But seeing my mother sitting on the floor waiting for me was painful; it was as if she didn't trust me and had to take her time to guard me like I was a rogue daughter. My girlfriend broke up with me at this time as well, so it felt like everything was caving in on me. I wanted my mother to trust me but she didn't; I wanted my girlfriend to love me but she left; I wanted to leave Saudi Arabia and make my life in another place but I couldn't come up with a plan.

This was during 2018, and a new round of reforms was being talked about. The government suggested it was going to lighten up, relax some of the suffocating rules, maybe even give some rights to women. I was skeptical from the get-go. The rulers of this country rely on a totalitarian absolute monarchy to stay in power; that's a hereditary dictatorship, governed along Islamic lines, that has never even considered accepting anything like the United Nations Declaration of Human Rights, which I'd been reading about on the illegal internet sites. These new, lightened-up rules might apply to places like Riyadh, but they certainly weren't likely to be adopted where I lived, in the most conservative part of the country. Actually, I really didn't believe things would change even in Riyadh. I had seen the consequences of liberal thinking in my country while surfing the internet. Many liberal thinkers had died because they were considered secular. I was eighteen years old and in my senior year of high school when I learned that forty people were hanged because they'd been heard speaking out about Islam. As students we were secretly texting each other to share the details we knew and cursing the denial of freedom of speech.

I admit that my last year of high school had been a challenge because my views on everything from my classmates to the school rules were hard to suppress. I had even switched to another school to make a fresh start. But my questions and curiosity came with me. One day I told my friends at school that I was going to run away; they thought it was a preposterous idea and said, "You'll find your father at the airport gate with a gun, waiting for you." I felt a rift developing between us, just as I had at the other school.

My tolerance for the criticism from some of the teachers who singled me out as a troublemaker was also in increasingly diminished supply. They spoke as though a good woman was infatuated with her husband, as if her role in life was to obey him; they claimed that Allah saw a woman leaving the home as abhorrent. I couldn't swallow any of this and felt they were evading the truth in all of our lives. Our teachers never spoke of the Saudi war in Yemen that was taking so many innocent lives. They never mentioned the crimes of some of the Saudi princes that everyone spoke about in whispers. And there wasn't a word about ISIS—the Islamic State of Iraq and Syria (also known as ISIL, the Islamic State of Iraq and the Levant). This terrorist group was very much on the minds of young people in Saudi Arabia because certain clerics were urging them to jihad, which means to fight against the enemies of Islam. So there was all this tormenting dialogue going on outside of the school and nothing but pious admonishments about the length of your skirt and the colour of your hijab inside. Or, in my case, the length of my hair, which was still too short for the liking of the principal. One day a teacher took my hand and led me to the side of the hallway and told me to start wearing clothes that were pink or red like most girls and to stop wearing clothes that looked like they belonged to men. I told her to leave me alone, but it led to another confrontation with the principal and another bawling out in front of the others. My position was clear: "I'm not a child, I don't feel like a boy no matter what you say about the colour of my clothing, and I didn't do anything wrong. Your standards of beauty and being female don't suit me."

I put up with the harassment because I needed to keep the peace at home and at school while I spent the evenings sharing my stories with my online group and absorbing the fascinating details they provided about how to escape. My grades at school were good, so that made my life at home easier. But there was a missing piece to my new secret life—it was my ex-girlfriend. Finally, I sent her a letter to ask how she'd been getting along since we broke up. Since moving to a new school, I had lost track of her. She responded immediately and we started talking and laughing and got together again. That's when we started putting together a real plan to leave Saudi Arabia.

I had an idea. It wasn't simple and would require patience, but I felt it could work. I had to convince my family to go abroad, and from there I would run away and seek asylum. My girlfriend's family travelled to foreign places, so I figured she could escape from one of them. The first obstacle I had to face was that my family didn't travel to Western countries; we only went to nearby Arab countries. But I had a plan for fixing that. I would have to plant my idea like a seed, give it time to germinate and then water it carefully and consistently until it was ready to bloom. Lamia and her husband had gone to Europe for their honeymoon. I began my scheme by asking Lamia to tell us about what she saw and how she liked it. She told us about the landscape—how green everything was compared to our desert land. She talked about the delicious food and the kind people they met in Bosnia. And she told us funny stories about how cold it was and what they did to stay warm. My mother was hanging on every word—it was working. But although it was obvious that my mother wanted to go, she was clearly afraid of being in a country with customs so different to ours. When I suggested we go there for a holiday, she said, "I am afraid for you! Your brother Majed will reject the idea for fear that a drunk Western man will rape you! Then Majed will kill you and kill the man, and he will also kill himself if something like that happens!" What a tirade in response to a trip to see something new! Her dismay made me think my escape plan was in trouble, so I suggested we go to Turkey instead. I was already imagining being in a beautiful new city and even thought about being by the seaside without my niqab, without my abaya. My mother interrupted my daydreaming and said, "Well, give me a chance to talk to your brother Mutlaq. Maybe he'll agree."

A few days went by while I worried about whether or not my mother would stick with the plan, but then she did it—she told Mutlaq that we wanted to see a foreign country. And to my everlasting amazement he agreed and began to put plans together. He even said he'd take care of half of the expenses for the trip.

Now I had to skirt around the timing. Ramadan was coming up at the end of May, and it took precedence over everything else. This was the third or fourth Ramadan that I did not fast (no food from dawn to dark), which is a requirement in Islam. Of course I would be beaten for failing to fast, but the violence in my life had become the norm and so no one paid much attention to it. This time, however, I pretended to fast because I wanted to avoid any consequences that could get in the way of my plan to convince the family to travel. Despite all this careful attention to keeping the family calm, my brother Fahad forgot to pray and didn't go to the mosque and ended up getting such a beating from Majed that I was afraid he'd need to go to the hospital. Majed came at him with an iron bar and struck him across the face until he was bleeding and unable to talk. It was as though he wanted to kill him. Even my mother was scared by this monstrous action. Religion is such a powerful weapon to Saudis, and especially to my older brothers. I see it more as a danger to our lives than a service to Allah.

For this Ramadan, I did every single thing I could to gain favour with my family: I treated everyone with the utmost kindness, and I cooked their favourite foods when we broke the fast at night and served them with love. I saw this as my last Ramadan and wanted it to be full of kindness and goodness. I wanted the abiding memory my family would have of me to be one of a loving daughter.

Eid al-Fitr came immediately after Ramadan. It is the feast to break the fast and my favourite event of the whole year. Everyone gathers early in the morning to pray the Eid prayers, which mark the cleansing of sin everyone has done by fasting. Then all the women and men of the neighbourhood go to the same mosque to pray. Even though I hadn't been a believer in a long time, I loved this tradition because everyone smiles and blesses each other whether they are strangers or old friends or family. Afterwards, we meet with our neighbours, usually at our house, and there we have the feast of Eid, which includes tables full of scrumptious food.

Soon after Eid, while I was still quietly encouraging my mother to get the holiday plan moving, I noticed that my older sister Lamia had started dressing differently. She'd always been the fashionable one in the family; in fact, she got away with bending the conservative wardrobe rules more than the other girls in our family. I figured it was because she was the eldest and had a good relationship with our mother. But now she started wearing really plain clothes, drab skirts and blouses, and told us her husband had started controlling her clothing choices. None of us said anything to her, but my heart ached to see a happy, well-dressed girl losing her style and image because of a man. As much as I felt sorry for her, I hoped with all my heart that this wouldn't be my destiny.

At last, Mutlaq announced in July that we would be travelling by plane to Turkey for our holiday. He also said that only four of us would go—Joud, my mother, Mutlaq and me. I wondered if this could be my chance to run away and decided to test my plan once we got there.

The trouble started as soon as we arrived in Trabzon, Turkey, in the middle of the night. When we entered the airport, wearing niqabs, the official asked my mother and I to remove our niqabs so he could see our faces. I knew my brother would be outraged by this and held my breath, waiting to see what would happen next. Mutlaq was so threatening that the man began to stammer and sweat, but he stuck to his rules and said if we wanted to enter the country the face coverings had to come off. I saw it as a delicious win and felt a certain pleasure in watching Mutlaq squirm in the face of a person with more authority than he had.

With that altercation behind us, we picked up the rental car and drove to the hotel. Everything was exciting to me—the beautiful city, the luxurious hotel, the thrill of being in a foreign place. Our rooms were in separate parts of the hotel, with my brother in one wing and my sister, mother and me in another. When my mother and sister slept, I would sneak out without my abaya and niqab and walk around the streets outside. What an adventure—I talked to strangers, drank coffee in the café and felt thoroughly entertained by everyone I met. I also started to think this might be the time to take off, to flee to Georgia, a country north and east of Turkey. My online group had discussed this plan with me before I left, suggesting that I cross the border and ask for asylum at the French embassy. One of my friends had done precisely that—she crossed into the capital city of Tbilisi and asked for asylum at the French embassy, so I figured I could do the same.

While I was marauding around at night, I was busy checking out maps to the border and bus stops and schedules and wondering how I could snatch my passport from my brother's bag. The people I talked to said the road to Georgia was precarious, known for bandits and the hijacking of cars. What's more, they said, it was a route through the mountaintops and very isolated. The more I heard, the less sure I was about leaving. In the end I realized it was not meant to be. I wasn't devastated by the lost opportunity; I felt I had learned a lot and tried out my escape wings, which would better prepare me for the next time. At least I had convinced my family to travel to a foreign country. Next time that part would be easy.

When we got home, I asked immediately when we might go again and reminded everyone about the wonderful time we'd had. My hopes were dashed when my mother and brother agreed that it would be two or three years before we'd take another trip like that. I fell into a prolonged silence, wondering what I could do and whether I should have taken the chance open to me while we were in Turkey. I was frustrated and unhappy, and I stayed that way until classes started at the University of Ha'il in September.

I had begged my parents to allow me to go to university in another city—hopefully Riyadh or Jeddah where the rules were slightly less conservative, where boys and girls talked to each other and where I could be away from my family and change my life—but they flat-out refused. So I registered at the University of Ha'il in the fall of 2018. Once inside the campus, we could doff our abayas, but what we wore underneath was not exactly provocative—long skirts and long-sleeved blouses, no makeup. If your hair was long, you could take off the hijab; if it was short, you had to cover it.

I couldn't even choose the courses I wanted to take, and that infuriated me. My mother and Majed used to discuss what I should study at university right in front of me. It was as though I had no opinion and no right to make decisions about my courses, as though they were discussing some other person—not the daughter, the sister, the human being sitting in the room with them. This kind of ostracism was driving my plans to escape. There was a lot of research I needed to do, a lot of very careful planning. My biggest fear at that time was that I would grow old in this place, that I would never be free, never realize my dreams.

Despite all of that, I didn't lose hope. I started school and presented myself as a new person, a quiet girl who didn't like mixing with others. Throughout the first semester, I even started to get closer to my family; it was as though I was trying to say goodbye to everyone. I was determined that this was the year I would start a new life away from Saudi Arabia—no matter how hard the task, no matter the cost to me.

Then my attention turned to Majed, who was bound and determined that I would withdraw as a student from the university and began to thwart me at every turn. Thankfully, my mother got involved and took my side. She was angry, which made me happy, and said in front of everyone, "When I leave this life, I want all my daughters to have a university certificate and a job." Then she said to me, "Men are jealous of women and they don't want to see them be more successful than themselves."

I went off to school with these mixed messages in my head. At last my mother was encouraging me to reach for my dream, but my dream was to escape Saudi Arabia, and I would spend all my spare time finding the key to the exit.

My girlfriend went abroad to study at the same time as I started college. It was an adjustment, to say the least: the university in Ha'il was as big as a small town, with dozens of buildings and beautiful gardens, as well as security women at every gate to search the students, check our bags and prevent us from leaving the campus during classes. There were strict rules that required girls to be on the campus all morning, even if we didn't have classes until noon. But my cohort was lucky because the year I began, the system changed and I was allowed to leave the campus as long as I showed my schedule at the gate to prove that I was not skipping a class. It was easy to falsify a schedule, and I saw that as a means to get out and be free. I took off every chance I had to go out with my friends or have a meal in a restaurant or coffee with other students who had also skipped by the guard at the gate. But mostly I went out on my own to be in the sunshine, to feel the air on my face. I'd doff my niqab and do normal things like sit in a park, eat ice cream at a hawker's stand, read books from the library or sometimes just walk down the street exploring the area and the shops.

It wasn't normal or acceptable for people in my city to see a girl without a niqab, so whenever I asked for a driver to take me where I wanted to go, he would ask if I was Saudi or not. When I answered that yes, I was, and that I was also from Ha'il, he would start hitting on me and telling me I looked different from the city girls! Nobody wants a girl on the street. Just as had happened to me before when I ventured out without a guardian, there were bullies and jerks who told me to cover my face and men who made filthy remarks to me about being unaccompanied and bare-faced.

In October 2018 I left the campus to go to the bookstore in town for textbooks that I needed. My next class was at 11 a.m. and I was worried about being there in time, so I hired a driver to take me back to the campus. I was in the back seat of the car as we drove to the school; he was quiet, didn't talk, but kept looking at me in the rear-view mirror. Then he turned around to look at me, as though he wasn't sure about what he had seen in the mirror. Obviously he was surprised that my face wasn't covered. He started asking me personal questions about whether I was married or had a boyfriend to go out with. I didn't answer, but I did laugh in a sarcastic way. Ha'il is surrounded by exits to the desert and the mountains; in a matter of minutes you can go from a busy street to an isolated road. The driver kept looking back at me and then, suddenly, we were on a road leading out of the city and into the mountains. He kept saying, "Just two minutes, just two minutes." I was freaking out and yelling my head off, but by now the car had pulled into the woods and stopped and he was crawling into the back seat and pinning me down. I begged him to let me walk away. He said to me, "There's no one here. You can't go anywhere." Then he raped me. I wasn't strong enough to fight him off. The whole time he was assaulting me he kept telling me to be calm. When he was finished with me, he simply crawled back into the front seat and drove the car to the university without saying a word. I used my phone to send a text to a friend to tell her where I was in case the monster at the wheel dropped me halfway to the city. When the car stopped at the school, she was there pulling me from the back seat. Both of us were crying while she admonished me for going out alone and for not covering my face. She was actually angry with me. We were yelling at each other so loudly that another girl came along and said, "Be careful, people can hear what you're saying and they'll start talking about this." My friend also reminded me that my family might find out if I didn't keep quiet. And she said, "Everybody knows your father." That was code for *Your life is at stake*.

There it was right in front of me—all the evidence I ever needed to understand where girls stand in Saudi. The driver who raped me simply drove away; he knew he would never have to be accountable for his crime. My friend at the university wanted to guarantee my silence so that my famous father and his family would never find out. Why? Because if anyone knew that I was soiled goods, I would have to be killed—it would be a classic case of honour killing. Of course the driver knew that. Of course my friend feared for my life more at the hands of my family than at the hands of the rapist. I was only a girl who wanted to feel the fresh air on her face. I was a target of male rage from a stranger and potentially from my own father.

When I got home, I felt weak, disgusted by my body, full of hatred for everyone who had hurt me or hadn't helped me. I thought about the escape plan that I'd failed to initiate in Turkey and then started blaming myself for everything in my life—the beatings, the ridicule, the rape, the refusals so numerous I couldn't even keep track of them.

It wasn't until I signed on to the secret social media site that night that I found the support I needed to get my fighting spirit back. One of the girls I was very fond of told us she'd been raped. Her words and then mine reminded us of our goal to get away from a life that treats girls like this. Later that month my best friend online managed to escape to Australia. Once she was safely there, I relaunched my dream of leaving Saudi Arabia and starting a new life in Australia. I'd read a lot about the country and thought it was the perfect place for me. I told my girlfriend about the plan and she said she was ready to leave everything and meet me in Australia. The thought of sitting in a café with her in Australia, going to beaches and swimming in the sea in a bathing suit and living in a place where girls were equal to boys was so joyful I could hardly wait to make it all happen. I was determined to make my dream come true, to finish my education in Australia, to become an actor. I uploaded apps and figured out where I wanted to live in Melbourne, and from that app I found new friends to talk to about life in Australia and told them I would move to Australia soon.

It was like weaving a tapestry that brought all the threads of a good life together.

Next I had to work on an exit. I started by chatting with my mother about how nice it would be to visit our relatives in Kuwait at the beginning of the new year. While she was pondering that idea, I used the tip I'd learned from the girls online to hook onto a friend's bank account, deposit my money into it and use that account whenever I needed to buy something I would need for my escape. Students in Saudi Arabia receive a stipend from the government every month. I had started collecting those funds and now deposited them in my account. I also asked my parents—each one separately—to give me money to buy various things that I said I needed. I deposited all of it. And my mother gave me an allowance each month that I added to the tally. Little by little the account grew. I felt so empowered by this; I was taking action and finding my way. When I had enough money, I applied for the Australian visa online and paid for it as soon as I was approved. It was like ticking off items on a shopping list—each tick brought me closer to my escape. I also realized I had to curb my instincts to cut classes, sneak out of the house, meet friends at forbidden cafés. Nothing must get in the way of my plan. I couldn't afford to make trouble that might result in me being grounded at home or expelled from school. I needed to cling to the Saudi state of affairs for the first time in my life.

In the midst of all this clandestine plotting, my sister Lamia gave birth to her first child—a darling, beautiful baby girl who found her way into my heart immediately. Every time I held her and looked into her eyes, I wondered what her mother would say to her about me after I'd gone. When she turned eighteen, would they tell why I ran away or keep the truth from her and tell her that her aunt Rahaf is dead?

In late November a fight with my brother Majed yanked me out of the reverie I was in following the arrival of my niece. It was over a dentist appointment. He took me there. I knew the dentist so I walked in, had the appointment and came back out to where my brother was waiting. Once we were a few feet away from the clinic Majed grabbed me and started strangling me. He was apoplectic, screaming, "Why did you walk in front of me and say hello to the doctor, a strange man, like I'm not a man with you?" I kept telling myself to calm down, take the beating, get to the other side of this quarrel without too many complications. But he had done the very thing I hated the most about my life in Saudi—he'd acted as though I was an inanimate object or invisible person, just because I'm female. I remember biting my tongue that day but promising myself that I would avenge my brother and everyone in my family who made me feel that I counted for nothing because I'm a woman.

In December, while I was preparing for final exams at university, the family began to talk about a holiday in early January. I held my breath while the plans went ahead and were clawed back and went ahead again. The decisions seemed to change every week. Who would go? Not Reem; she wasn't well enough to travel. Lamia of course had to stay with her baby and husband. All of my brothers said they would go. That scared me, as it would make my escape trickier. At last it was decided: we would leave December 31. It was just a few weeks away.

The conclusion came easily to me. It was time to go. I turned to the network and shared my decision and asked them to help me hatch the final stage of my plan to flee. The family vacation to Kuwait was set. Now I needed to work on the details—the precise escape tools I would need. Discussing the plan with the network helped me to know what I had to do. Then, very carefully, I turned each piece of the plan into a blueprint. Every step required trust in the network, a bit of luck on my part and more patience than I had ever practised in my whole life. But step by diligent and exacting step, the plan began to come together. I had the visa and money from the undetected bank account—enough to buy a plane ticket and pay my way during what could be several days of hiding and dodging authorities.

I was ready to go.

CHAPTER SIX

Escape

{BANGKOK: SUVARNABHUMI AIRPORT
SATURDAY, JANUARY 5, 9 P.M.}

When I realized I had walked into a trap, my sense of freedom and relief vanished. Reality crushed my happy dream of escaping the abuse of my family, the injustice of my country and a future that meant being tethered in marriage to an old man who would have absolute control of my life until I died. Being murdered by my father and brothers for daring to run away was another truth that I knew I would face. How else could I explain the early-morning departure from the hotel in Kuwait while my family slept, or the ticket I'd purchased for the flight from Kuwait to Bangkok, or the valid Australian tourist visa stored in my phone? My plan was so carefully prepared; I'd stay a few days in Bangkok at a hotel I had already booked and then fly to Melbourne, where another Saudi runaway would meet me. I'd ask for asylum as soon as I made it to the other side of the arrivals area, and my new life would begin.

Instead I was trapped here in the airport. My fear was like

a hot white light exploding in my brain. I was shaking and sweating; I could see my future becoming unravelled, my own life ending. The people in front of me—the man I'd foolishly presumed was being kind to me, the woman who wouldn't look at a terrified girl—they froze as images in a horror movie while my brain slipped into slow motion and the past and present collided. As much as I felt like an animal who'd been caught in car headlights, the survivor in me knew I had to calm down and figure out what to do next. I was quite sure they were going to arrest me and send me back. I needed to figure out how to get in the way of that plan.

The Thai man, who turned out to be an agent of the Saudi embassy and who I now considered the enemy, told me to sit down on a chair near the passport control counter. My passport was in his hands. I was trying to decide between begging him to help me and attacking him so I could run into the main part of the airport. He clamped his hand on my arm and led me to the chair, where I sat for fifteen minutes that felt like fifteen hours. I was scanning the room around me, wondering which person might help. While I sat there the face of Dina Ali came to my mind. She's a Saudi girl who ran away, just as I did. She was trying to get to Australia but was stopped in the Philippines, just the way I was stopped here. The official took her passport as well as her phone because her ticket was on it. She spoke to another woman in the airport, a Canadian tourist, and told her she was in danger, that her family would kill her, and asked if she could use the woman's phone. Between them they contacted every organization they could think of—human rights organizations, humanitarian organizations, the Manila police, the Manila newspapers—but they couldn't get through. The officials at the airport said they had received a call from a very important person in Saudi Arabia who told them to hold her documents and not allow her to leave. She posted a message on the internet that read:

> *My name is Dina Ali and I'm a Saudi woman who fled Saudi Arabia to Australia to seek asylum. I stopped in Philippines for transit. They took my passport and block me for 13 hours just because I am a Saudi woman. With the collaboration of the Saudi Embassy, if my family come, they will kill me. If I go back to Saudi Arabia I will be dead. Please help me. I'm recording this video to help me and know that I'm real and here.*

The Philippines airport officials said they had called her family, and sure enough two men turned up at the airport saying they were her uncles. And that they would take care of her. She was screaming in the airport, saying that the man who was insisting to officials that he was her family and had come to take her home was not her father; she was begging airport officials and tourists in the airport to help her. But no one helped her. The uncles left and a seemingly sympathetic lawyer arrived and told her not to worry, that he would get her passport and documents back and she should come with him. The Canadian woman who had stuck with her throughout this ordeal figured Dina was safe and bade her farewell. But this was a total trick. The so-called lawyer was a Saudi handler. The incredible thing is that no one in these airports understands the peril a runaway girl from Saudi Arabia is in. Everyone in the world knows that the status of women and girls in Saudi is life-threatening. The international community knows what goes on—I read their reports on illegal internet sites. Heads of state are aware as well. So why isn't there a hotline to call? Why is it airport officials have the power to hand a girl over to certain death? I had seen the video footage of what happened to Dina Ali next when I was still at home in Ha'il, looking at social media sites. A passenger in the departure lounge recorded the shocking abduction of an innocent woman. They had taped her ankles together, placed tape over her mouth, tied her to a wheelchair and thrown a blanket over her while she struggled and squirmed and was pushed through the departure lounge and onto a Saudi plane. To this day no one has heard from her and no one knows where she is.

I figured I was about to face the same fate. I sat in the chair, wondering what would happen next, what I could do to save myself. Then I saw a collection of men strutting toward me. The Thai man and five security guards, along with another man who looked Arabic, were walking toward me like a gang of executioners. Even before they reached me, I knew what had happened: my family had awakened, realized I was on the run, called my father, and he had used his power and influence to call the airport authorities and instruct them to set a trap for me. I thought, *This is it; my life will end, they're going to put handcuffs on me, they're going to take me.*

They were walking to where I was sitting—I stood up and stepped backwards. I didn't speak a word, but I'm sure the immense apprehension and almost paralyzing fear I felt were showing on my face as I backed away from them, still trying to concoct a getaway plan. That's when they told me the truth: there was an alert from the Saudi embassy about me, submitted by my family. The Thai man said he was there as a mediator for the Saudi embassy and it was his job to take me back home. The Arabic-looking man, who was actually

Kuwaiti, said he was working for the Kuwaiti airline. I knew instantly that I wanted to expose them; I wanted someone to know what was happening, because this was inhumane. A young girl, obviously terrified, was about to be set upon by seven men. I felt I was in the hands of a Mafia gang.

I was also remembering what the Saudis did to Jamal Khashoggi. He was the Saudi Arabian dissident who dared to speak out against the crown prince and the king about the lack of human rights in the country. I followed his story on the illegal internet sites where I learned everything else about my country. Khashoggi was constantly criticizing Saudi Arabia in the column he wrote for the *Washington Post* and during his regular appearances on the Al-Arab news channel, and he had to leave his homeland in fear for his life the year before I bolted. He was getting married and needed documents for the ceremony and went to the Saudi embassy in Istanbul to get them. He was never seen again. Later it was proven that he was murdered inside the embassy; his body was dismembered and removed from the embassy by a team of hit men who flew from the kingdom to Istanbul to assassinate him. A year later, just six weeks before I fled, I read on social media that even though the royal family admitted he'd been killed, no one had been held responsible. I remember seeing his face on my phone and thinking what a friendly, smiling man he was, and there were photos of the woman he was going to marry; she was waiting outside the embassy, patiently waiting for the man she loved who never came back out the door. I was taken by that photo—it was caught on the security camera. She looked so alone, probably happy and full of plans for the future at first, but later . . . I wonder how long she stood there before sounding the alarm. I looked at that photo when

I read about his murder and thought, *Her happiness is dashed.* But even the worry she must have been feeling was nothing compared to the horrific truth she would eventually know.

Standing there staring down the gang of thugs in front of me, I wondered if I would end up like Jamal Khashoggi or Dina Ali, kidnapped and killed or disappeared.

The men stood there together, watching me, speaking to each other behind their hands. I couldn't hear what they were saying, but I decided to videotape them while recording my own voice saying, "The embassy has stopped me." I sent the video to a friend of mine. Although I didn't have time to post everything, at least I got that much out so someone would know where I was and what had happened. I was trying to put on a confident face in front of these men, but the truth is I was nearly hysterical at this point. I didn't know what I should do. In all the careful planning I'd done, there was no backup calculation for this. I contemplated running away, jumping over the customs barrier where my chair was and dashing into the main part of the airport and asking for help from a crew member of a foreign airline—a Western airline—thinking they would be humanitarians. What should I do? Was the future really out of my hands? Submitting to them seemed like a bad idea. My suitcase was still in checked baggage. The Thai man had my passport. I had my backpack. I decided to run. I had no clue where I was going but hoped I would find an exit and get to the city, where I could disappear. The Thai man came toward me. I ran at him as hard as I could and pushed him backwards with one hand and tried to grab my passport from him with my other hand. I got it. Immediately I hid it in my jeans because I knew he wouldn't dare touch me to try to get it. The security guards had anticipated the direction I would go and hopped over the barrier to cut me off. They were coming at me from both sides now, so I stopped running. The Thai agent looked nervous and the other six men gathered around me, forming a circle that I knew I could not penetrate.

Once I was surrounded, the Kuwaiti man said something that astonished me: "I wasn't expecting you to be wearing clothes like this when the plane landed." So, he had been watching for me, monitoring the footage on a camera at arrivals, looking for a woman wearing an abaya and niqab. "I thought you would look like a Saudi Arabian woman," he said. I thought to myself, *Why would he be monitoring the camera, and how many people does he follow like this?* Then he told me, "You look normal. Why did your dad tell me you are sick and need treatment for mental illness?" I was stunned, absolutely shaken by this remark, and asked him, "Did my father say that?" He said, "Yes," and he showed me a text on WhatsApp that my dad had sent him—it was a faked patient folder from the mental hospital saying I was mentally ill. And it included my photo so I would be easily recognized as soon as I arrived in the airport. Suddenly I froze. Goosebumps popped up all over my body. I felt everything around me go dark, and I remembered what my dad had said about my sister Reem the night she tried to run away. At the time, none of us had understood what she was accusing our father of. How could the father of a girl commit such a terrible atrocity? We all agreed that poor Reem was suffering from an anxiety that we didn't know about. The details of that terrible night and the days that followed were flooding my mind as I absorbed what this Kuwaiti man was telling me. When Reem tried to tell my family the awful things our father had allegedly done to her, he said she was mentally ill; he sent her to the psychiatric hospital and she was silenced by drugs that wrecked her mind. When she came home, she was like a zombie. We'd needed to take care of her ever since. Standing there in the airport I suddenly understood what Reem had said that night. What my father had done. I also now believed that by claiming she was mentally ill and sending her to the psychiatric hospital, he had effectively condemned her to a life of dependency on mind-destroying drugs. I even began to wonder if those in charge of her care had been paid off. Shame, shame, shame on this man who believes in the code of honour.

I composed myself and tried to convince the Kuwaiti man that my father was powerful enough to put me in an institution and ruin my life. It was obvious this man sympathized with me and I was certain he didn't believe what my father said. So I couldn't understand why he didn't want to help me. Was he just another man who believed girls were expendable and that there was nothing he could do to change that ancient fact? I felt helpless, caught in the sinister hands of powerful people who have authority. And at that moment I surrendered and gave up my plan to get away; my dream to live freely was in ruins. I stood there feeling dismayed, thinking, *It's over. I'm being sent back.* I sat down in the chair, picturing my return home and preparing myself for the reality that my life would end. I felt sure my father would kill me. I wondered if he would kill me right away or hide me from the world where no one would hear about me or find me ever again. I wanted to cry but for some reason tears did not come. I don't know how I remained so strong on the outside despite the turmoil inside me. My mind was flashing to the future, to the consequences I would pay for my attempted escape; I was even sitting there on the airport chair hoping that after I vanished I wouldn't be forgotten. I wanted someone to know my story.

After a while, I don't know how long, I stopped caring about my failed plan and fell into a fog. I wasn't focusing on anything when I heard words that turned everything around. The Thai man said the trip back to Kuwait would be in two days and they would put me in a hotel here in the airport to wait for that flight. What? Two days? A hotel? I could hardly believe my ears—and I actually thought, *THERE IS STILL HOPE*. I was sure this was a sign to fight for my life and my freedom. Instead of figuring out the next step, I went on instinct and decided to make a run for it again. I was running in all directions, trying to get away from those men and shouting, "Help me" to anyone who would listen. I turned on my phone and recorded my pleas and the responses from tourists, shopkeepers, airport officials who kept looking at me as though I was acting out or being a typical teenager or simply a problem best ignored. Here's a recording with one airport official:

ME: Please help me.
OFFICIAL: Your visa is not granted.
ME: But I'm in danger.
OFFICIAL: Yeah? How?
ME: Yes, it is so dangerous to me.
OFFICIAL: What do you mean "so dangerous to you"?
ME: Saudi Arabia is dangerous for me, so I can't go back.
OFFICIAL: You have to wait for the Kuwait plane. It does not go to Saudi Arabia. You came here on a Kuwait plane so you will leave on a Kuwait plane. You can't stay here.

The conversation was ludicrous. But it was evidence that just as the airport officials in the Philippines refused to help Dina Ali, this guy had no interest in helping me.

That was the end of the conversation because the security goon had caught up with me, pushed me backwards so hard he hurt my chest, and then dragged me back to where I'd been sitting while I kept yelling for help. People noticed me, looked at me, but nobody helped me. Nobody. Then I was taken to an office in the lower level of the airport. I was watching carefully, looking for a way out. I saw a door that led to the outside of the airport, to the street. The door was open; I could see the sidewalk, the road and the sky. How could I get myself through that door? I kept staring at it, trying to memorize the location, recording the images on either side of the door, hoping I could find a way to ditch my minders and get back there and make a run for it. The Kuwaiti man saw me and laughed when he said, "Don't even think of running away. If you embarrass the Saudi embassy, you will regret it." A veiled threat. I thought again of Jamal Khashoggi.

Once in the office on the lower level of the airport, I made a strategic error. They asked me to sign papers in the Thai language and to show them my passport as I signed them because they needed additional information. Maybe my fatigue level was surfacing or maybe I was preoccupied, but I let my guard down and pulled the document from the pocket in my jeans. Like a coiled-up snake making a strike, the Thai man suddenly leapt toward me and snatched my passport and tucked it away in his pocket. The look of triumph on his face was pitiful to me—here was a grown man tricking a scared girl in a foreign place and rendering her helpless. He reinforced everything I had learned about men:

they will do anything to anyone to save face, to look like winners in front of other men when in fact they have broken the laws of decency.

We left the office, walking like a marching band of seven men and one young girl into the main halls of the airport, which were filled with shops and washrooms and crowds of people carrying their luggage. I asked the Thai man to give me back my passport. He said he would give it back to me in two days when my flight was ready for departure and I was in the boarding lounge. As we walked along, the Kuwaiti man started acting like we were old friends, chatting with me and telling me that my cousin was a friend of his and they all work at the airport. I was trying to compute this information and wondered if this was how it worked—that my father had called my cousin and the string of commands and threats went through him, so the Kuwait Airways people feared retribution from my father and, rather than doing their job to protect their passengers, they capitulated and did my father's bidding.

I was still trying to put all that together when he said, "Your father is going to call me. Do you want to talk to him?" I said, "No, never." He said, "What about your older brother?" I was crying at this point and agreed that I would talk to Mutlaq. When I heard his voice on the phone I told him, "I'm sorry. I swear all I wanted and planned for was to leave in peace without anyone noticing." He said "Why?" I told him, "I cannot explain because you're not going to understand me or side with me. My problem is with the Saudi regulations and laws, not with our family." He said, "Okay, you're going to live in Kuwait after this." I told him, "I don't believe what you're saying. For three years I was asking both you and Mom to at least allow me to study in a different city than our city, and you refused, so how are you going to allow me to live outside Saudi Arabia?" I was sure he was lying; this brother Mutlaq who studies in a different city, he would likely do anything at all to capture me and drag me back to my fate. But family ties are powerful. So even though Mutlaq had been despicably mean to me, even cruel, hearing his familiar voice on the phone when I was being held by strange people in a foreign land really wrenched my heart. Now I was crying so hard that I could hardly speak. My brother seemed emotional as well; his voice started to sound as though he was about to cry himself. He promised me that nothing would happen to me and said he would be waiting with everybody to welcome me in the Kuwait airport when my flight landed. I asked, "Who is everybody?" He said, "Me, your father and the Saudi embassy." I figured this was another trap and said, "Are you kidding? Why is the embassy going to welcome me?" He said, "For your safety." Whatever smidgen of trust I had in my brother evaporated instantly. I disconnected the phone and decided I couldn't trust anybody.

By now I was so emotionally distraught that my nose started bleeding all over my clothes. I remember the looks of the people passing by; they seemed to be scared and avoided making eye contact with me while I was crying, gulping, gasping and holding my shirt to my nose to stop the bleeding. *Why don't they stop and ask if I need help? How can they look the other way? What's wrong with people that they are so busy minding their own business they can witness a kidnapping and fail to act?* The Kuwaiti man rushed to my side and asked me about the blood. He thought I was trying to harm myself. I said, "It's just a nosebleed."

We arrived at the airport hotel—aptly named the Miracle Transit Hotel—and I was told I would stay here until my departure. The Kuwaiti guy pointed to one of the security men and said he would be on guard in the lobby to make sure I didn't try to run away. With that warning he handed me the key to my room and pointed to the hall I should follow to get there. He and the Thai agent and another man I didn't recognize from the Bangkok airport were all in the lobby, watching me while I walked to my room. I saw them because I was examining every nook and cranny, even looking over my shoulder to where they were standing, checking every door, trying to find an exit I could run to. I opened the door to the room slowly, entered, and started trying to figure out if it was safe or not. The wall was glass from floor to ceiling; I could look out and see the interior of the airport but not outside. I thought how scary this place was, how easy it would be for them to incarcerate me here. I felt trapped. Even if I screamed, no one would be able to hear me. I thought, *I can break that glass and get out; on the other hand, someone could break it and get in.*

I sat in the room and started tweeting about my situation—where I was, what was happening. "I am the girl who escaped Kuwait to Thailand," I wrote. "My life is at stake and I am now in real danger if I am forced to return to Saudi Arabia." At first, I hesitated to publish my photo and full name. I was tweeting to my runaway friends in Australia, Canada and Sweden; one of them said I should post a video of myself and say my name so the world would believe this was real, that it was really happening. I had to think about the consequences of doing that. I dropped offline for two hours. Although my absence from Twitter made everyone worry, I needed to get hold of myself. I had to decide whether or not to post my picture. This was a very big decision. I was sitting on the floor of the hotel room. My heart was beating very fast and I was breathing fast, too fast—panting, gasping for breath and trembling—and I wanted to cry. I was really scared of posting my name and photo because of how my people back home—my family, friends, relatives and neighbours would react. I was also afraid that posting my name and face would mean I'd lose freedom in another way. If people know who you are, there are restrictions on your life, people expect things of you. I wanted to be free and safe.

Many people lent their help by amplifying my voice online. Mona Eltahawy, an Egyptian American journalist and social activist based in New York City, saw my tweets in Arabic and translated them into English. Along with Mona, it was my three friends from the secret network that stuck with me throughout my ordeal in Kuwait and then in Bangkok who convinced me I needed to take a risk and show my face on social media. They basically said, "Say your name and show your face or you're going to die."

I thought about the consequences and finally decided that the world should know I exist; if anyone was going to help me, they needed to know who and where I was. So I made the decision to publish my picture and full name.

I discovered almost immediately that my networkers were right—that action got me the attention I needed from human rights groups and the media. Mona Eltahawy continued to translate my tweets and told the Twitter world how much trouble I was in. She tweeted, "Rahaf's father is a governor in #Saudi Arabia. He is a powerful man. She has requested asylum and fears for her life. She has said she fears her family will kill her if she is returned to Saudi Arabia."

I also posted this video on Twitter: "My name is Rahaf Mohammed. I am 18 years old. I can't do anything because they have my passport, and tomorrow they will force me to go back to Kuwait. . . . Please help me. They will kill me."

Twitter lit up immediately with responses from all over the world.

The Saudis really know how to let their world standing dip to the lowest of the low. Twitter has done some great things this week—and highlighting your plight is one of them. Without your phone, the truth could not have been told. Keep going Rahaf.

They don't care. It's never had consequences before and it likely won't have any real consequences now. The west is still more than happy to arm them and hopelessly addicted to their oil. Nobody in power seems willing to truly confront them.

So any future survivors trying to escape should take note and have spare phones hidden on their person, under their clothes or in luggage somethere [*sic*] safe . . .

No, better to have people waiting in transit airports that you can trust.

Announcements such as this make the citizen lose confidence in any government entity and in any embassy and consulate, as if they were one gang working against the citizen. [translated from Arabic]

Saudi Arabia is basically Gilead. I wish you all the best and that many others have the chance to follow you to a better life. The UK government should hang its heads in shame for it's [sic] support of this regime.

You know what disturbs me: I only see men in the video, talking about a woman, telling what she should do. #SaveRahaf

This is horrible, but you're bringing attention to the awful behavior of others and shedding light on it for the entire world to see. You're making a difference not in your own life, but in the lives of other women. Stay safe.

You're a smart girl Rahaf! Stay ahead of the game...always a couple of steps ahead...

I also received a lot of attacks from people in Saudi Arabia cursing me when I declared I was giving up Islam. They wanted to stop me, wanted me dead. Even people outside Saudi Arabia in the Arabic world who were Muslim threatened to kill me because I was trying to break the shackles for women looking for a better life in a better society.

Many journalists from the *New York Times* and from Sydney, Australia, and the UK contacted me, and I gave them the details about being held like a prisoner in a hotel room in Bangkok so that my situation would be spread in the Western world as well. I posted the essence of my story—they put me in a room, they took my passport, my father will kill me, I want to get away. I posted all that and also said, "I'm going to stop eating until I find help and UNHCR [United Nations

High Commissioner for Refugees] comes here to help me." I was more frightened than ever before in my life. All I could think was, *What if they break down the door and enter my room and kidnap me because I was talking on social media? What if my outburst online and not being silent angered them?* I didn't only take on my family, I took on the authorities too. They were the people who could ruin my life.

With all this running like a tickertape through my mind, I decided I had to try again—to leave the room, to ask for help or to try to run away. I left my room barefoot so I could run faster and went into the airport area and found the guard who was supposed to be watching me—he was asleep. I left my backpack in the room but had my phone. I walked to the arrival area and followed the passengers, hoping I could follow them out of the airport without having to pass through a checkpoint like customs, hoping I could get away before someone spotted me. I even thought of breaking a window to get out. But nothing was working; I could not find a way. I knew it was an impossible and crazy idea but I had to try something, and so in desperation I once again turned on the recorder on my phone and went from one uniformed official to the next, asking for help; I even went up to tourists and told them my story and asked them to help me. I asked shop-keepers too, everyone I came across. I knew no one had the authority to help me; most of them couldn't even understand a scared, barefoot, crying girl explaining she's going to die if she is returned to Kuwait. People made excuses. They walked away from me. I found an airport employee, someone who would understand me; he told me I should return as ordered to my country. I insisted he call the Thai police. He refused and told me to go back to my room and wait for my flight.

Everyone in this whole airport seemed to know enough about my story to stop me from getting help. My hopes were diminishing fast. I thought maybe I could find a hiding place in the airport so that when the plane left, it would leave without me. I roamed around the hotel looking everywhere, but I could not find a way out that wasn't guarded. I even tried to remember the location of that room in the lower level, but all the images I'd tried to record in my mind about where it was eluded me.

After a while I felt my search mission was starting to look like the actions of an escapee and attracting the sort of attention I didn't want. So finally I decided to go back to the room, past the sleeping guard. But even in the room, I didn't feel safe. There were people knocking at my door telling me I should eat. That just jacked up my fear; I thought, *What if they want to poison me?* I decided to post the videos of me talking, actually begging people in the airport to help me, and of their sometimes nasty, sometimes dumbfounded responses, and also the looks on peoples' faces as they heard me and clearly wanted nothing to do with me and quickly looked the other way. The knocking at my door was becoming incessant—new voices asking me to go to their office so we could talk. Who could this be and what did they want with me? Could I dare to trust them? Were they trying to help me? There was always a sliver of faith, tiny but present, that made me believe good people would come to my rescue. But not the ones in this hotel. I decided the door knockers were menacing creeps on the same side as the gang of thugs who put me in this room and that my life was better protected on the inside. But as the knocking continued, so did my sense of terror. I could still hear the Kuwaiti man telling me that if I tried to leave the airport, the Saudi embassy would make me regret it. Then, as if on cue, the Kuwaiti man came to the door and said, "Open the door so I can talk to you." I shouted at him through the door and told him to go away and tell everyone to leave me alone.

By now exhaustion was seeping into my mind as well as my body. I hadn't slept in two days or eaten since I was at my aunt's house in Kuwait. I was scared and worried that this overwhelming fatigue would make me sleep. And it occurred to me that if I fell asleep, they could enter the room and kidnap me, and I wouldn't have the energy or power to fight or to defend myself or to run away from them. I stayed in that room for hours, listening to the harassing knocks at the door. Finally they stopped and I presumed the men outside went away.

I fought the overwhelming urge to sleep; there were only twenty-four hours until my flight would depart. I tried to think of a way to save myself, some exit or doorway or place I had overlooked, but increasingly I was consumed with thoughts about how my life would end. Would I wait for them to come and get me, or should I kill myself right here—break the mirror in the washroom and cut my wrist or set fire to the place with the lighter I had with me and take us all down in flames? I cannot find words to describe the feeling in your soul when you know your fate and know the end of your life is near. I was consumed with the facts of the lives of girls and women— we are half the world's population but we mean nothing to the people who run the world. I thought only Saudi Arabia and a few other Muslim countries were like that, but here in Thailand the men cared nothing for my safety, my truth, my life. What is it about being a girl that makes people think it's their right to control you, speak for you, plan your life and beat you whenever they feel like it?

I was falling ever deeper into despair when I received a

message on WhatsApp from an Australian journalist who'd been following my tweets. Her name was Sophie McNeill. She said she had contacted Human Rights Watch and Amnesty International in Bangkok and Sydney and sent them the tweets that were out there and asked if they could help me. And she wondered if I knew anyone else—maybe in Australia—that she should contact to help me. I had to tell her I didn't know anyone like that. She emailed others—journalists and UN refugee agency people. It was when she got a reply from the Human Rights Watch deputy director in Asia, based in Bangkok, that she told me help might be on the way. She said he spoke Thai, knew people at the Bangkok UNHCR office and even had good contacts at the Thai government office. Right away he started tweeting about me and said he was "extremely worried that [a] Saudi woman Rahaf Mohammed al-Qunun will face similar fate [to Dina Ali] if she is forced back from #Thailand. She wants to seek asylum, currently being kept at #Bangkok airport hotel by representatives of #SaudiArabia embassy." Sophie thought this was significant for my case; she said now it wasn't just Saudi activists and friends tweeting about me. If Human Rights Watch was worried about my fate and publicly willing to lobby for me, that would help me to gain the attention of the UN and foreign embassies in Thailand. Then Sophie sent me a text to tell me to hold on. "I'm coming to Bangkok," she said.

I clung on to those words like one who's dying of thirst and is offered water, and waited the rest of the day and all night hoping there was at last a rescue. Did I dare believe that someone would fly from Australia to Bangkok to know my story? Could I trust her? I had to—she was my last hope.

I didn't move from my room until I heard from Sophie when she arrived in Bangkok at 4:30 in the morning and contacted me to ask the best way to meet without raising too many alarms. She wanted to know if anyone was watching me, anyone outside my room. I gave her the name of the hotel and its location in the airport and hoped we could meet in the lobby, but it wasn't possible to do that without attracting attention.

Sophie stayed in the lobby to keep watch, to make sure no one could take me. She suggested I get some sleep, but I don't think I slept at all. Sometime after 6 a.m. Sophie informed me that human rights advocates said I must make a claim for asylum right away, at the hotel. A colleague of Sophie's, who worked for the Australian Broadcasting Corporation in Bangkok, had arrived by then and suggested I make my refugee claim with their cameras rolling.

I walked toward the lobby, trying to pump up my level of confidence, feeling scared but somehow triumphant, as though the door to my future might be opening. The minute my eyes met Sophie's, I felt safe. I even started smiling despite my turmoil and kept smiling because waves of relief were rolling over me. At last there was someone on my side who knew what was happening to me.

I walked over to a hotel employee and asked her to call the authorities so I could ask for asylum in Thailand before the flight I was to take to Kuwait would leave. She ignored me. It was as though I was invisible. I tried another airport official—same confounding brush-off. Sophie started tweeting about what she was witnessing now that she was at the hotel. She posted a tweet saying, "There are guards outside the hotel room. It's 6:20 a.m. in Thailand. She's been threatened to be put on the 11:15 a.m. Kuwait Airways flight. She's been denied access to a lawyer. She wants to speak to UNHCR Thailand and claim asylum." Then I started to hear from people suggesting I defend myself by going limp and playing dead or screaming my head off and kicking anyone who came near me.

We kept at it, trying to get someone to help me, until shortly after 8:00 a.m., when a Thai immigration official appeared. He wouldn't listen to my demand to file a claim. Eventually I figured I'd better get back to my room and lock the door because the security guards were coming at nine to take me to the departure lounge for the 11:15 a.m. flight.

Not long after, Sophie slipped by the security guard and joined me in my hotel room. I started worrying that the authorities would come to my room and use a hotel key to open it, so I decided to barricade the door. Sophie was recording my pleas for help and sending my voice out to the world, telling people I was seeking asylum. She also recorded me dragging furniture to block the door so that even if the officials had a key, they wouldn't be able to get in. First I pushed and pulled and finally got a table about three metres by two metres to the door. It was really heavy. Despite my exhaustion I kept pushing it until I had it positioned against the door, but it was a haphazard protection, to say the least. I wondered, *Is this going to work out and will this save me?* I decided the table wouldn't be enough of a barrier, so I cleared everything off the bed, upended the mattress, slid it on its side to the door and crammed it in beside the table. All I could think of now was putting as much stuff as I could against the door so that no one could enter. I even put a chair on top of the table, even though I knew very well that chair wouldn't make a difference.

At 9 a.m. I was ready. The door was blocked with piles of

furniture, my plea to the United Nations had gone out with my name and my face and my words begging for help: "I am not leaving my room until I see UNHCR. I want asylum."

As frightening as the situation was, it wasn't without its moments of humour. One man sent me a tweet that said if I gave him US$20,000 cash he would come and rescue me in the airport. Imagine that! But there were issues that were becoming increasingly problematic. Since I was in a hotel within the airport, no one could get to me unless they had purchased an airline ticket, which would allow them to enter the secure area where the Miracle Hotel is located. But Sophie told me that as the story about my dilemma spread around the world, the media were turning up and that many had managed to get into the departures area. Some of them were even waiting at the Kuwait Airways gate and ready to record whatever happened if I was put on that flight.

By now people were posting on Twitter with information that was not true. A television program called *Yahala* was interviewing a Saudi official from the Bangkok embassy who denied that my passport had been taken from me. This was a lie, but he confirmed that the Saudi embassy had called Thai authorities at the request of my father after he found out that I had escaped. Then another distortion: the Saudi official said that I did not have a valid visa to Australia. Yes, I did. I told Sophie about an ominous text I'd been sent saying that people who knew my father were trying to get me back. I felt sure my father was behind it. He was rich and had power, and everyone standing with him would do as he said. I never underestimated the reach he would have in getting me back.

People started coming to my door. The exchanges that followed, with the Kuwait Airways employee, the Thai man and some compliant woman they brought to my door, could be seen as comical if my life hadn't been at stake. They dodged between demanding, begging and cajoling. I was tweeting the details as fast as I could. Sophie was recording all of it. The Kuwait Airways employee was actually the person working with the Saudi embassy to help confiscate my passport.

The Kuwaiti representative said, "Open the door. What's wrong with you?" I said, "I can't." He said, "Can you open the door just a bit?" I didn't reply. A while later, a woman knocked and asked if I'd like any breakfast. We had launched into a good-cop, bad-cop scenario. Then a Thai official knocked and said I had to leave. "You don't have asylum in this country," he said. "You cannot take asylum in Thailand."

I'd done everything I could. Now they were knocking again, telling me to open up, it was time for my flight. Where was UNHCR? Was I not a valued-enough human to get their attention, or did they presume, as all these hotel people did, that I was a rebellious teenager who should be sent home? I knew that on the other side of that door was either my death or a brand new beginning. The knocking stopped. There was an eerie silence at the door. I was checking my phone for the time every two or three minutes. The flight departure was still an hour and forty-five minutes away. What was happening? Like in a chess game, I was waiting for the rook to make a move. Sophie and I talked about what might happen next. I was reading the tweets that were coming in—hundreds of them from around the world kept pinging up on my phone. The heartfelt messages let me know there were a lot of people out there cheering for me, hoping I would be rescued. They were addressing me and Kuwait Airways and the UN. Some of these tweets read:

We are with you.
Is this happening right now?

Keep fighting for your rights and freedom.

Stay strong.

Save Rahaf.

Do not allow your airline to deport her.

Time is running out.

This teenager is doomed.

Yell scream let everyone around know you are HERE.

Her family will kill her it is inevitable.

The whole world is worried about her.

We hear you, don't go anywhere.

This is a death or life case.

Stay strong we are praying for News.

I hope for a happy ending.

She wants to seek asylum.

God bless her.

I read her tweets it is horrifying I'm so sorry for you Rahaf. In solidarity with you.

I tweeted back:

based on the 1951 Convention and the 1967 Protocol, I'm rahaf mohmed [*sic*], formally seeking a refugee status to any country that would protect me from getting harmed or killed due to leaving my religion and torture from my family.

And again:

I seek protection in particular from the following country Canada/United States/Australia/the United kingdom, I ask any [of its] Representatives to contact me.

I was still obsessively checking the time on my phone. Sophie was recording; I was tweeting. She was connected to reporters on the outside. They were monitoring the flight I was supposed to be on. I was actually lying down, totally exhausted and trying not to sleep, when Sophie said, "Rahaf, the flight has gone." I sat up and asked her to say it again. She said, "There is another flight to Saudi Arabia in fifteen minutes but the flight you were booked on has left." It was a victory for me. I was incredibly relieved and immediately fell into a deep sleep.

In the meantime, Sophie heard from her colleague that UNHCR officials had been at the airport for three hours but had been denied access to me by the Thai authorities. Soon the UNHCR put out a statement that said, "UNHCR has been following developments closely and immediately sought access from the Thai authorities to meet with Ms. Mohammed Al-qunun . . . to assess her need for international protection. . . . UNHCR consistently advocates that refugees and asylum seekers—having been confirmed or claimed to be in need of international protection—cannot be returned to their countries of origin according to the principle of non-refoulement . . . [which] prevents states from expelling or returning persons to a territory where their life or freedom would be threatened. This principle is recognized as customary international law, and is also enshrined in Thailand's other treaty obligations."

I slept for three hours, and when I woke up it was to news I could hardly believe: Sophie's contacts said the Thai authorities were processing my request for asylum. We waited inside my room in this Miracle Transit Hotel. If hope could be measured by heartbeats, I had a tally that could make wishes come true.

Around 4 p.m., a woman was at the door knocking and saying, "Excuse me, madam, the UN is here. We will not send you back to your country. Don't be worried." But Sophie's contact at Human Rights Watch told us, "Don't believe it. Wait for the UN." I logged on to my Australian visa account and saw that it had been cancelled. I stared at that revoked document and thought, *This is truly what despair looks like.*

Who would have cancelled my legitimate visa to Australia? At first I wondered how far my father's power reached. Could

Saudi Arabia actually instruct another country like Australia to cancel a visa? What sort of deal with the devil do the officials in two countries make when they basically push the sacrifice button? But I reconsidered and realized my conclusion was wrong. It probably wasn't my father's power that cancelled my visa. More likely it was the Australian government that invalidated it because they knew I was going to seek asylum. I'd heard from my friends in Australia that this is how they treat Arabs. I already knew some of them had been deported upon arrival at the airport because the officials there don't want to deal with asylum cases. In October 2018, they treated my Saudi friend very badly and asked to speak to her father to make sure she was allowed to travel. That's how I knew I should have the name and contact details of a man who would pose as my father in case they asked to call him when I arrived.

The runaways I was in touch with were in Sweden, the UK, Germany, Canada and Australia. Once I'd made my decision to escape Saudi Arabia, I had chosen Australia. I suppose proximity matters, but my choice was mostly based on all the good things I'd heard about it. I knew it was a good country and that I would be able to go to school, work and live my life without beatings and fear. Women have rights in Australia. There are laws that say if you beat a woman or girl you will be punished. My friend lives there and says the people are kind, the beaches are beautiful and we can do everything there that we cannot do in Saudi Arabia.

A while later, a tweet popped onto my screen from the global head of media for the UN refugee operations. It said, "Dear

Rahaf, my @refugee colleagues are at the airport now and are seeking access to you!"

At 5:57 p.m. there was another knock on the door. I crawled over my barricade to look through the peephole. "Who is it?" I asked. "The UN" was the reply. "Proof, show ID, I need proof," I said. They did. A business card slid under the door. It had the familiar UN logo and the name of one of the people I could see through the peephole on the door. They were asking me to let them come in. There was a semi-circle of Thai soldiers and others who were probably UN people around them. I pushed the piles of furniture out of the way and I opened the door.

The UN people said immediately they would protect me. They also told Sophie she had to leave, which scared me because I trusted her and felt safe with her. Then they asked me about the reason I tried to escape, about my life and my family. They recorded everything I said and took my photo and told me they would do everything in their power to make sure nothing happened to me. They said they were taking me to a hotel in a secret location where there would be very tight security protection while they examined the documents I would need to establish a refugee claim. They returned my passport to me and I showed them the valid Australian tourist visa on my phone that had now been cancelled. They had already launched formal arrangements for my long-term asylum status.

Then we left the room that had been my refuge for the longest two days of my whole life.

As we walked toward the airport exit that I hadn't been able to find, the Thai official who was overseeing my immigration status walked beside me, saying, "We will not send anyone to die. We will not do that. We will adhere to human rights under the rule of law." I thought, *Where the heck were you when I begged for help from all those people in the airport?* There was a crush of security, UN officials, airport authorities and media around me; it was as though I was walking in place on a moving ramp cocooned inside a circle of men. When we exited the airport, despite the entourage around me I felt the setting sun on my face and thought immediately of my exit from the hotel in Kuwait at 5 a.m. two days earlier. The soft wind and warm sun that had touched my face and neck as well as my soul that morning was still there, thousands of miles away, assuring me that I too would rise again.

The UN checked me into the Royal Princess Larn Luang hotel in downtown Bangkok and came back later with my checked bag, which they had retrieved from the arrivals carousel. Sophie stayed there too but wasn't allowed to be with me. I even asked her to buy me some cigarettes but the UN said she could not deliver anything to me. There was no television in the hotel room, but I used my phone to check the news and was astonished to find my story on every broadcast. My face was plastered all over the place. I was wearing the jeans with blood on them from the nosebleed I'd had earlier, the same short-sleeved shirt I'd been wearing, and a look on my face that taught me a lesson: here I was, in the middle of what had become an international news story pitting Saudi Arabia against the United Nations, walking out of a death trap and into daylight and safety, but my face did not show the terror and anxiety and fatigue I had felt. Instead I saw the face of a girl who was determined, strategic and maybe slightly surprised.

I read a tweet from the UNHCR that said, "Thai authorities have granted UNHCR access to Saudi national, Rahaf

Mohammed Al-qunun, at Bangkok airport to assess her need for international refugee protection. . . . For reasons of confidentiality and protection, we will not be in a position to comment on the details of the meeting."

Once again, I presumed the drama had come to an end. I lingered in a long, hot shower and prepared to fall into bed hoping for a peaceful sleep and knowing that the UN staff would meet me in the lobby of this hotel in the morning. Little did I know the third chilling act of this drama was about to begin.

Two UN officials came to my room before I turned out the light. They told me that my father and my big brother Mutlaq had just arrived in Bangkok; they had gone to the Thai authorities to ask to see me but had been refused. My father called the UN and said that if he couldn't see me, he demanded the right to talk to me over the phone. The UN man asked me if I wanted to talk to my dad but advised me to refuse the request. I told them, "I don't need any advice. I wouldn't even consider talking to my dad." And I warned them about my father's influence on others and his ability to have his own way. By now my alarm system was fully operational again.

Actually, I felt certain that my father and my brother had come there to kill me. I tried to let the UN people know that I was scared that someone would help them. I knew how things worked in Saudi Arabia, so even though there were police in front of my door and in the lobby of the hotel, I still didn't really feel safe. I even found myself checking the lock on the door and wondering if my father and brother could find their way to my room, force their way in and kidnap me. And it occurred to me that my father could try to bribe these policemen to open the door and let him come into this room. Or what if these guards came in and took me to him? I put a table against my door—just like in the hotel room in the airport. I was even too scared to sit on the balcony to smoke, thinking they might be outside and waiting for the chance to shoot me. There's no limit to what they can do when their so-called honour has been smeared. There were UN men staying in the room next to my room and in other rooms as well; I don't remember how many there were. But they were good and they protected me every step of the way.

In the morning the UN men took me to a hospital to have a checkup as well as blood tests to make sure I was healthy and hadn't been injured. Then we went to the Australian embassy for a meeting with immigration and refugee officials there. But most of the day was spent in my room, where I was on my phone watching the unfolding melodrama of my own life even as it was interrupted by advertisements and weather reports. I heard reporters talking about how there was pressure on Australia to take me. And every channel I clicked on was featuring the arrival of my father and brother and the backstory about the lives of women and girls in Saudi Arabia. Most of them featured footage of poor Dina Ali begging for her life in a Philippines airport. And all of them spoke of the guardian system in Saudi Arabia that gives a father, husband or son the right to control everything a mother, daughter, wife or sister does throughout her life from birth to death. They described my life perfectly: I'm not allowed to marry; I'm not allowed to get a job or leave the house or travel anywhere without the permission of my guardian. Women are treated like minors even if they are fifty or sixty years old. A teenage boy can beat his grandmother for sitting in the garden without permission. I wanted to shout at the videos I was watching, "That's what has got to change in Saudi Arabia. That's what drove me away."

I followed the shifting status of my situation all day on the Al Jazeera television network. The UNHCR representative in Thailand said, "It could take several days to process the case and determine next steps. We are very grateful that the Thai authorities did not send her back against her will and are extending protection to her." Oh yeah? They hadn't protected me until the UNHCR came along!

Thailand's immigration police chief was also interviewed. He said, "The Kingdom of Saudi Arabia has not asked for her extradition." The embassy tweeted that they considered this issue a "family matter." Then why did they send a Saudi official to pose as an airport employee and claim he was helping me when I landed in Bangkok?

One report after another quoted sources who really were trying to help me. Human Rights Watch called on Australian officials to allow me into the country. Their Australian director said that since Australia had expressed concern in the past about women's rights in Saudi Arabia, it should "come forward and offer protection for this young woman."

The Australian government claimed it was monitoring the case closely and that it was "deeply concerning" that I had said I would be harmed if I was returned to Saudi Arabia. A senator called on her government to issue me an emergency travel document so I could fly to Australia to seek asylum.

Al Jazeera said, "It . . . comes at a time when Riyadh is facing mounting pressure over the killing of Saudi journalist Jamal Khashoggi at its consulate in Istanbul in October, and over the humanitarian consequences of its devastating war in Yemen."

I also checked the UNHCR website. It said that refugee status is normally granted by governments, but the UNHCR can grant it where states are "unable or unwilling to do so." It also said that they do not comment on individual cases. One media outlet simply concluded, "Now that Ms Mohammed al-Qunun has been given this status, another country must agree to take her in." But I listened to another comment by the deputy Asia director of Human Rights Watch, who said he was worried about me. "She said very clearly that she has suffered both physical and psychological abuse. She said she has made a decision to renounce Islam. And I knew once she said that, she is in serious trouble."

I posted a new message on Twitter saying, "Don't let anyone break your wings, you're free. Fight and get your RIGHTS!"

One of the reporters caught the Saudi chargé d'affaires in Bangkok on a hot mic while at a meeting with airport officials and Ali, the Kuwaiti guy who was on to my case like a mosquito. The chargé d'affaires had no idea the microphone was on when he said, "They should have [taken] her phone instead of her passport." The men at the table laughed like co-conspirators. Even with the international community watching, even with the world knowing what happens to girls like me in Saudi, this man seemed to think that I should have been captured and returned to my fate at the hands of my father and brothers. And the men at the table either agreed with him or didn't have the courage to dismiss his disgustingly guilty comment.

As the day wore on, my story continued to headline the news. During the afternoon, the Australian foreign minister arrived from Sydney and was scrummed by reporters on the steps of the Australian embassy. I was hanging on to every word she said, but to be honest I was expecting to hear her say that she'd travelled here to process my asylum papers and take me back to Australia. Wrong. The way she replied to the reporters' questions put me right back in escape limbo—*Am I safe? Will I make it? What are the UN and government people saying to each other?* The foreign minister said that after talking with Thai officials, they were assessing my claim for asylum. Then she said, "There are a number of steps still to be taken in the assessment process." By now I was gulping and going into panic mode. *Oh my god, does this mean they aren't going to take me after all?* When the reporters pushed her as though interpreting my own questions, she went on to say, "There are, as I have just said, a number of steps in the process, including in terms of that assessment. They are required to be taken and they will be completed within due course and then that matter will be resolved." She also said there was no time frame and that there was no possibility I would be going back with her to Australia. She said I would have to get in line like everyone else.

I can hardly describe the way I felt at that moment. I wanted to shout at her, "I've been waiting in line my whole life, being mistreated and beaten and feeling I was dodging death. I have risked everything, including my life, to escape the injustice of a country that clearly hates women. Which part of this do you not understand?"

A long day turned into night. The television reports intensified. So did my anxiety. What I didn't know that night was that the UNHCR team was working behind the scenes to ensure my safety. They needed to fast-track my asylum application, mostly because after the Thai police informed them that my father and brother were in Bangkok looking for me, they were increasingly worried about my security. The next morning the men from the UN came to my hotel and took me to the Canadian embassy. The meeting was short and sweet: the ambassador said, "Would you like to live in Canada?" I said, "Yes." He said, "Your visa will be ready at three o'clock this afternoon. You are booked on a flight to Toronto tonight." In two short sentences my fortunes changed. I hardly knew what to say. Thank you seemed inadequate. I felt tears welling up in my eyes as I tried to convey my gratitude to this man and to the UN people who were with me.

Later I learned that the UN had used what are known as back channels through the night. They felt the Australians were taking too long with their decision. I also found out that the deputy director from Human Rights Watch in Bangkok had been "working his contacts," as they say in their business, to get me safely out of Thailand. He said Canada offered immediate asylum, something Australia was unwilling to do. I understand he said, "She needed to get out of Thailand very quickly. Her brother and father were still here and these were the people she feared. The Thai government was also very keen to have her moved as quickly as possible and the feeling was also that Saudi Arabia is a very influential government and it has a lot of capacity to pursue people, particularly women who could tarnish their image if allowed to remain free on an international stage."

I was on a Korean Air flight that departed Bangkok at 11:37 p.m., bound for Canada.

As soon as the plane was airborne, UNHCR issued a statement saying I had left for Canada. They explained that they'd had growing concerns about my safety and about the uncertain timelines from the Australian government about whether or not they would grant me asylum; they had referred my case to Canada, and the Canadian government had processed my application in a matter of hours.

Arriving at Pearson International Airport in Toronto felt more like a lift-off to a new life than a landing. I'd been told by the Canadian ambassador in Thailand that someone would meet my plane—he might have even said who that would be, but as the plane taxied to the gate I had no idea what was next. The flight attendant came to my seat and said I would be taken off the plane first. Three airport officials were at the top of the ramp and said someone from the Canadian government was waiting to meet me in a nearby office. I didn't know who was who in the government of Canada, but I soon realized they must care about me, because the person waiting to greet me was the minister of foreign affairs, Chrystia Freeland, and her secretary. But more than that, the minister had brought her daughter, who is my age, with her, which made me feel very welcome. Her daughter gave me a hoodie with *Canada* written on it. I had a ball cap from UNHCR, so that's what I was wearing when I left the arrivals area. Two women from an organization that helps refugees settle in Canada joined us a few minutes later and explained they were the ones who would help me find a place to live, register at a school and buy some warm clothing. It was winter in Canada and I noticed the cold right away. The foreign minister assured me that it would get warmer. The women also warned me that along with the media there were crowds of well-wishers; they would speak to them to say I had arrived safely. I felt I should go with them. As the doors to the exit slid open, I could hardly believe my eyes: a crowd of reporters, photographers and TV camera crews, as well as dozens of people I didn't know, had

assembled in the arrivals area. As soon as they saw us coming
through the door they began clapping, cheering and shouting
"Welcome to Canada" to me. I could hardly process what I
was seeing. Having grown up in a place where women have to
be covered and hidden and regarded as invisible, I was being
greeted as though I was somebody who existed, who had the
right to be there.

Minister Freeland told the crowd to quiet down and said,
"This is Rahaf al-Qunun, a very brave new Canadian." She
had her arm through mine but let go of me to step toward
the crowd and catch a bouquet of roses someone was throw-
ing to me. I was trying to absorb this: a government official
meeting my plane instead of turning me back, then catching
flowers in mid-air and tucking them into my arms. Believe
me—this was so far from my experience, I was watching
it as though I had wandered onto a movie set. Then she
told the crowd, "It's obvious that the oppression of women is
not a problem that can be resolved in a day, but rather than
cursing the darkness we believe in lighting a single candle.
Where we can save a single woman, a single person, that's a
good thing to do."

As much as I had struggled for this moment, I couldn't
believe it was actually happening. I couldn't speak as I was
overwhelmed with the sense that I was free, that I was born
anew; it washed over me like an aura that made me feel giddy
with happiness. I felt loved and welcomed. I had an enormous
rush of pride and gratitude—my voice had been heard, these
people understood. Freedom is the most important thing for
a person. I gave up everything to be free.

As we left the airport I went back to Twitter, the social
media platform that I had relied on throughout this odys-

sey, because I wanted to say thank you to all the people who helped me, believed in me and took this long, scary ride with me. I tapped out my message:

> I would like to thank you people for supporting me and saving my life. Truly I have never dreamed of this love and support. You are the spark that would motivate me to be a better person.

Then I leaned back in the car and wondered what the next chapter of my life would bring.

Chapter Seven

Triumphs and Consequences

I opened my eyes in the early morning of the first day of my new life to three pieces of news: there were one hundred death threats on my phone; my father had disowned me; and a snowstorm was about to hit Toronto, the city where I was launching the new me as a Canadian. I hadn't come this far to be undone by such barriers. I shut down my social media account, dropped my family name al Qunun and went out to find a store where I could buy a parka that would keep me warm.

The death threats had begun while I was in Bangkok—furious postings from Saudis who were outraged that I would dare to run away, leave my religion and my family. They came mostly from my own tribe, from people I had never met but who felt I had stained our lineage and desecrated their reputation. And, in Saudi fashion, they wanted revenge. Most called for a public hanging; some suggested I be whipped to death. Other threats came from Muslims in Arab and non-Arab countries warning me to return to Islam or face death at their hands.

As I scrolled though Twitter I read:

I will pay someone a few thousand dollars just to kill you

we will follow you to kill you

I swear to God that I will let you suffer before I cut your head

you deserve death and I will be the one who sends you to Allah

One thing I noticed immediately was that all the threats, every single one, came from men. There were no women threatening me. As I scrolled through the tweets from the safety of my new home, I wondered again what it is about the men and boys in Saudi that makes them think these hateful thoughts. Before, I examined this as an insider who was frightened about the consequences of coming under the scrutiny of a father or brother or, heaven forbid, the mutaween. But this time I was on the outside, and my observations as a free woman were much more troublesome. This was misogyny, the hatred of women, and it's sanctioned by the government. How does it continue—the guardianship laws, the disgraceful gender apartheid inside and outside the home, the dismissal of a woman's voice? I have to conclude that it is fear that drives these men—fear of women; that, if given the rights other humans have, Saudi women are seen by Saudi men as competition for jobs, for rank, for the very virility men cling to like a lifeline. I would like to have an honest discussion with a man who thinks that as a fifty-year-old it is okay for him to marry a twelve-year-old. Or a man who thinks having four wives is really, honestly something Allah wants him to do, and that it has nothing to do with his view of male power.

The death threats worried me, of course, but they also made me lament my homeland, a country that sees its mothers, wives, sisters and daughters as evil Jezebels that the men need to be protected from, even though their foolish pretense is that we are fragile creatures that need their protection.

After deleting the death threats, I moved on to the news from my family, who had released a statement claiming they had disowned me. It read: "We are the family of Mohammed El Qanun in Saudi Arabia. We disavow the so-called 'Rahaf Al-Qanun' the mentally unstable daughter who has displayed insulting and disgraceful behaviour." Those were familiar words—*mentally unstable*—the ones my father used to describe Reem when she explained what my father had done. I was actually surprised at the size of the hurt I felt reading those words. Not the mentally ill part—that was just a ruse my father was using—it was the "disavow" part that squeezed my heart. I love my family—even my older brothers, who beat me so severely; even my mother, who rarely took my side but did give me sound advice about making sure a man never had control over me; and even my father, who I always thought would take my side if he was ever at home long enough to know what my side was. And I thought about my sweet little brother, Fahad, and the absolutely adorable Joud, and of course Nourah Mom. Did they all disown me? Did they vow not to be related to me ever again? The statement from the family—a.k.a. my father—made me cry, but it didn't stop me from following my dream for one minute. That's why I dropped the family name al Qunun and decided henceforth to be known as Rahaf Mohammed.

There are several runaways from Saudi Arabia living in Toronto, and just like my online family that got me here, they became my survival guide, telling me where to buy a parka and boots to deal with this startling event called winter. I could see my breath when walking outside, and the sidewalks were slick with hardened snow that challenged every step I took at first. But the trees laden with soft puffs of snow were like a drawing I could only have imagined before. Icicles hung from branches and caught the light, refracting colour that dazzled the artist in me, and as much as the frosty air nearly made me gasp, there was something about this winter wonderland that had an appeal to me, as though I was a pioneer in my new life. However, as much as I embraced the changes, I was continually drawn—almost daily at first—to my Saudi side. That ingrained second skin Saudis wear, the one that says that girls don't count and girls should be silent and invisible, struck me like car brakes when I walked outside alone, when I met my friends at a café, even when I shopped for a pair of winter boots.

My runaway friends took me to their favourite cafés, and we often talked about news from Saudi. For example, while I was in flight mode, the mutaween struck again, this time ferociously. We read the story online from the *New Straits Times* claiming the religious police had arrested more than two hundred people for supposedly violating public decency. Viewing this news from a different and safe place, I saw it even more clearly for what it was—harassment and that

ever wilful need to punish women and girls, this time for
so-called inappropriate clothing, which translates to colourful abayas and less suffocating niqabs. The *New Strait Times*
ran a photo online of a woman with a headband and a face
mask. I thought it was clever, but the mutaween said it was
a sinful act that showed too much of her pretty eyes. They
claimed people dressing like this were offending public
morals. And this was happening just after Crown Prince
Mohammed bin Salman vowed to loosen the ridiculous
ancient restrictions. Tourist visas were issued for the first
time and the ban on cinemas was cancelled; so was the ban
on women driving cars. And for those who live in Riyadh
(certainly not Ha'il), concerts and sporting events were
allowing women to enter. However, it was at a concert—the
Beast Music Festival in Riyadh—that another eighty-eight
attendees were arrested for wearing immodest clothes (make
that tight-fitting clothes, or clothes with so-called profane
language or images) and also for public displays of affection.
My friends at home texted me to say they were all afraid the
religious police were on the prowl again.

Those early days in Toronto also allowed me time to
unpack the events that occurred in Bangkok and find out
how truly fortunate I was to escape. It had seemed like an
eternity while I waited minute by minute in that hotel room,
concentrating on staying alive; now I could examine the complicated details that happened around me during that desperate forty-eight-hour period. Imagine the reporter Sophie
McNeill buying a plane ticket and flying from Sydney to
Bangkok to get my story. That's what reporters do; I understand that. But I also learned that Sophie insisted on covering
this story because she knew the Dina Ali story. She knew that

there should have been some form of protection in airports for girls like me, or anyone who is in mortal danger. Putting the airport authorities under a blazing media light was the way to find justice. She knew that the best hope I had was the Twitter campaign coupled with human rights organizations that could sound the alarm and basically scare the Thais into letting me go. She had to clear a lot of barriers to get to me, not the least of which was a nine-hour flight from Australia to Thailand, but she was determined.

I also learned more about what the UNHCR people and government officials refer to as back channels. The people at UNHCR were on the phone all over the world, talking to human rights organizations and governments, trying to make a plan to get me safely to another country. While I was in Bangkok—once the UNHCR rescued me—they worked with lawyers in Thailand to file an injunction preventing my forced deportation. I also learned that Thailand's chief of immigration police admitted that authorities in his country had acted at the behest of Saudi Arabia. There was a lot I didn't know until later—some of it shocking. I knew the UN people were worried because my father and brother were in the city, making threats and using their Saudi power to get what they wanted. And I easily imagined how concerned they were, because I know what my father can do. But I didn't know the depth of their concern until later. They said the reason they approached Canada to take me in was that Australia was taking so long to make a decision, and with my father and brother in town, they wondered if they could in fact protect me. I also found out that the wretched man who pretended he was helping me when he asked for my passport as I arrived in the airport was really an agent sent by the Saudi embassy to stop me. How does a Saudi agent get into the secure area of an airport? And how is it that airport staff are told to knock on my hotel room door and lie to me—I can still hear that Thai woman's voice saying I could stay in Bangkok and didn't have to worry because the UN was here. How many governments allow their officials to lie and pose as the UN, using criminal behaviour in an airport full of foreigners?

I also learned that the number of asylum seekers from Saudi Arabia on the UNHCR radar has quadrupled in the last five years. Most claim asylum in the United States. Canada is the second choice. And the networkers' website says there's an increase in people seeking help on their site since my escape to Canada was in world headlines. An exodus. I hoped it would fuel the fire in the anti-guardianship law protest I had secretly signed.

But as much as these facts occupied my mind, I was more caught up with learning to live in a new place and in an entirely different way. While I felt that I had been reborn, my abiding fear in those early days was that my family would find me, that I would disappear and no one would know what happened to me after that.

Running away is not easy, especially for a girl my age who has no life experience. I had to learn how to do most things that girls in Canada do without a second thought—going to a shop without supervision, trying on clothes, paying for my purchases (my brother always did this). I hardly knew what to say and worried about making mistakes. If you have never lived under guardianship, you cannot know how restricted a person can be. In Canada, young women my age do their banking privately and as a matter of course. I was not only unfamiliar with the currency, I had no idea how to bank using automated teller machines. I began to realize that Saudi Arabia was still inside me and I needed to find a way to cut the customs and obligations that were tailing my every step. In the early months, I felt afraid to enjoy the freedom I had, reluctant to take steps on my own, the very steps I'd run away to take. I hesitated to go out to buy coffee, so I asked the settlement counsellor from the organization that helps refugees to go with me and to do the talking and even the purchasing. It wasn't because the currency was new to me or that I didn't know what I wanted. It was because my brother's voice was playing in my mind. The way you wouldn't walk into a burning building even if you wanted to rescue someone, I felt the censorship and control I'd been raised with holding me back from the other side of the world.

As though extrasensory perception was controlling the Saudi part of my mind, I continually felt I was being watched, judged, exposing myself to harm by doing something forbidden, even though I had defied my brothers often and easily at home. Walking alone on the street was part of the pleasure of being free, having the sun shine on my face, the wind blow my hair, even the snowflakes fall on my cheeks was a celebration of liberation. But those Saudi laws were stamped on my soul, so I had to actually grow accustomed to being free. That same feeling came over me when I went to the market to buy food or to the bank to open an account or to the hospital to have a checkup. I was like a child with no experience in life or in doing things on my own.

My daily life was so different in Canada I could have been on another planet, but one thing was certain: I liked this place and knew I would become acclimatized quickly; I was finding my voice and learning to use it efficiently. With help from the refugee settlement organization, I was offered a room in the home of a Jewish family. It was a good decision, as I learned their way of life simply by being at the dinner table and watching the goings-on in the household. In Saudi Arabia we have our meals around a cloth called the dastarkhān on the floor; we sit cross-legged and eat with our hands. But this family ate with knives and forks and spoons and sat at a table. I had to learn to use the cutlery, and angle myself at the table so that I could manage this new custom. I felt awkward at first, especially as the family watched me as though I was from a different planet, but I learned to eat just as they did. Living with Canadians was the quickest way to acclimatize to this new country. For example, the family was hockey crazy—as I found out so many Canadians are. They watched all the games on television, talked about the players as if they were part of the family and the complicated rankings of the teams—if this team beats that team then the other team becomes the one to beat—as though it was a game of mathematics. I quickly learned the rules of this game that was played on ice at breakneck speed, and although I never attended a game and still haven't learned to skate, I became a fan of the home team—the Toronto Maple Leafs. It was good for me to live in that house because I was part of the everyday habits and lifestyles of Canadians, but eventually it was better for me to move on. I found a room in a hostel where many refugees and immigrants stay and took the next step toward independent living.

Although it was not easy, one of the lessons I learned is that freedom requires work—it means making choices, living with the consequences of your choices, finding your way, righting the wrongs when you make a mistake. I was like a butterfly that had just burst out of its cocoon and was flapping its wings frantically for the first time, not knowing how long it would take me to soar.

In the beginning, I spent a lot of time by myself in my room trying to figure this out, asking myself what I was afraid of and why the Saudi rules that I hated kept seeping their way into my thoughts. I began to realize that freedom is more than avoiding a beating or being able to do what you want to do. It's being mentally and psychologically free as well as physically free. That was a pretty big leap in my thinking. It led me to decide I needed help sorting out these demons from the past that kept creeping into my new life. If Sophie McNeill played a major role in helping me in Bangkok, the people I relied on in Toronto came from the refugee settlement organization that was on hand from my very first moments in the city, after I was greeted at Pearson International Airport by the minister of foreign affairs and before I exited to a very welcoming crowd of supporters who made me feel loved and protected. This team had ideas and answers and tips I could not have imagined. They helped me find a place to live and guided me through the banking and shopping I would need to do. As for my education, the founder of a private school came forward to become another welcoming port in the storm I was going through. I was introduced to him by Chrystia Freeland, who was concerned about me completing my education. He took me under his protective and generous wing and invited me to join the classes at his school in Toronto.

But above all of that, these settlement counsellors understand the trauma a frightened newcomer encounters, even one like me who risked her life to get here. I asked them how to cope, and soon after, I started attending therapy sessions that focused on my thinking as a Saudi daughter and a Canadian newcomer.

By the time I'd had a few months of therapy sessions, the warm weather had arrived and I witnessed the rebirth of spring that Canadians embrace after a long cold winter and saw my own thinking blossom along with the flowers. I ventured out into the city, found new friends and learned a lot about myself.

I had to finally accept that the past is part of my future, that I need to find a way to weave my memories—even the dreadful ones—into the life I have chosen for myself. There was so much that was unfamiliar—the language, the behaviour, the socializing, even the laws. I was absorbing the new as much as I was sorting the old. In the process I felt I was unleashing skills I didn't even know I had. As it turns out, I pick up languages easily, and I can navigate the subway system, the bus routes and find my way without a brother leashing me in.

Not only that—this was a city like no other place I had ever been in. I had never imagined so many races and religions in one place, not to mention the huge numbers of immigrants and refugees in Toronto. A mix like this is unheard of in Saudi Arabia. They call it multiculturalism here, and everyone seems to be very proud of the fact that Toronto is one of the most multicultural and multiracial cities in the world. There's proof everywhere you go, from the subway to the streets, the shops, the restaurants; I hear there are more than 250 ethnicities and 170 languages spoken just in this city. And about half the citizens are from a visible minority—Asian, Black, Latin American, Arab. In fact, one chart I read said there are sixteen countries that have over 50,000 people represented in the

Toronto region, including 337,000 from India, 300,000 from China and 200,000 from the Philippines. I had no idea before coming here that I was joining what I now see as a worldwide experiment in multiculturalism, and I liked it immediately. In Saudi Arabia other ethnicities and cultures are frowned upon—in fact rejected outright. We tend to be very closed, maybe because we don't learn anything about other places in school or maybe because the government doesn't want Saudi citizens exposed to any other lifestyle. In Toronto, multiculturalism *is* a lifestyle; it's part of everything—food, fashion, art, music and city events. I was impressed right from the beginning by the way the people here take all this for granted. While I was whirling my head around to examine something I had never seen before or asking questions about foods I'd never heard of in restaurants, I knew I was becoming part of the fabric that weaves a multicultural city together.

I had such a fervent mixture of hope and fear, of excitement and occasionally despair, I knew I was a work in progress. For example, a lot of people stopped me on the street and told me, "You're the girl in the news." That was exciting, reaffirming and it made me feel like a celebrity, but it also created anxiety. It suggested I was abnormal, not fitting in, when all I wanted was to be a normal girl walking down the street minding my own business, rather than a peculiarity or an unusual object noticed by others. Having said that, I have to add this: people were very good to me; they never hesitated to say they were proud of me, some even wanted to hug me! It felt very reassuring to know that people were on my side in this quarrel I had with my family, my country and even a country like Thailand. It's natural to seek approval, so I really enjoyed that gift of acceptance and affection they were giving me. But anonymity was my goal and I hoped that soon enough I would become a natural part of the Toronto landscape—just another girl living her life, rather than "that Saudi girl runaway."

At first I was very attracted to doing things that were forbidden in Saudi Arabia—drinking alcohol, going to nightclubs and wearing shorts. I remember on my nineteenth birthday, just two months after I arrived, I was with my best friends in a restaurant and I ordered a glass of red wine. Suddenly I felt I had made a mistake, that I should not be drinking wine, even though everyone else was. That was the Saudi fear still living inside me.

Toronto is known as a city of neighbourhoods, so getting around the city was important as I had friends in one neighbourhood and events I wanted to go to in another, and I was attending school in still another part of the city. So I became a regular on what everyone calls the TTC—that's the Toronto Transit Commission. The intricate subway and streetcar system became my byway as well as an education of its own while I watched the people around me as though I had wandered onto a movie set. So many different faces, so many languages; my trips to school were an ongoing source of entertainment and allowed me to permeate the outer layers of this new place as if by osmosis, and to feel, little by little, as though I belonged.

In the meantime, the smear campaign from Saudi Arabia was relentless. Some said I was miserable and homesick in Canada. Others claimed I would be waiting tables for drunkards at Toronto bars. I was called everything from a drug addict to a whore by the Saudi media. But I knew where that was coming from. I had busted the Saudi code of conduct, but I was free of that wretched guardianship law and could now live my own life. The main newspaper in Riyadh, called *Al Riyadh*, used my escape to claim the need for families to protect their daughters from such dangerous ideas and to stop the brainwashing I must have been subjected to by some unknown assailant. People who helped me, such as Mona Eltahawy, were viciously attacked on social media in Saudi. She was no stranger to attacks from woman-hating men. I was becoming immune to it as well.

I discovered the high-octane power of a Twitter campaign, in particular the one I conducted from my hotel room in the Bangkok airport, when I learned the European Saudi Organisation for Human Rights, which documents and promotes human rights in Saudi Arabia, enlisted a lawyer to defend me in Bangkok against deportation back to Saudi Arabia. He said my tweets played an overwhelming role in preventing my deportation and that once the Thai authorities understood the strength of the international support my tweets were getting, their attitude changed completely.

I don't know whether the authorities understood the strength of the decision I had already made while I was detained in the Miracle Transit Hotel in the Bangkok airport. I decided that I would not be sent home, that I would end my life first. I wrote a goodbye letter to my closest friends and told them to publish it if I was forced back to Saudi Arabia. I still have that letter to remind me about the level of desperation I was experiencing, but I have tucked it away with some of the other memorabilia that's best kept in the past.

I knew early on that I was winning the refugee battle—finding my way, learning the language, figuring out the currency, making new friends—but no matter how hard I tried to reject the past and no matter how much time went by, I still had regular jolts like lightning bolts from my family that covered the gamut of emotions—rage, heartbreak and worry. For example, in March, I called my mother via WhatsApp to see how she was doing. In truth I wanted to hear her voice; she is my mother, after all. I hoped to know about the goings-on of our family, including my beloved grandmother Nourah Mom, who I worried about. I hoped that by reaching out to my mother I would repair the wound my sudden exit had opened, that she would be happy to hear from me and would want to know how I was managing. I especially wanted to know about Joud—the little sister I last saw when I was preparing to leave the hotel room in Kuwait. I'd looked at her—sleeping peacefully that early morning—and tried to record the image of her on the back of my eyelids so she would stay with me forever. I loved that little girl with all my heart. My mother answered her phone, and as soon as she heard my voice she started shouting—cursing my friends and saying they were the ones who changed me and brainwashed me with their insane ideas. She told me to go immediately to the Saudi embassy in Canada, to hand myself over to them and beg them to take me to Saudi Arabia. I tried to help her to understand that living in Canada was my choice and that I was happy here and hoped she would keep in touch with me and share news about the family. She disconnected the phone and then immediately texted me a message with the voice of my little sister crying and saying, "I love you, Rahaf." She knows that Joud is my weak point; she was using that girl's tears to make me return to a country that would probably see me dead even before they let me comfort Joud. I wept bitter tears afterwards and worried about the way they might be using Joud to trap me into returning. I had learned about the many burdens of freedom but also the empowerment it gives. I dried my tears and deleted my mother's phone number.

Later that month, my girlfriend from Saudi came to Toronto and we were reunited. It seemed to many like the denouement of a romantic novel—one where we had found each other in the chaos and confines of a country that would have had each of us killed for being in a lesbian relationship. Experiencing the freedom to be together on the street in Toronto and holding hands in public and staying together was like a dream come true. At first I was giddy with joy— this was the life I'd been searching for. We were even planning to get married. Then as time went by and the illicit part of the relationship turned into normalcy, the loving bond between us dimmed and our love affair turned into a friendship. By the end of May we had gone our separate ways, and although (like so many romances) we tried to be partners again during the summer months, it was not to be. By October we had parted once more, but we remain friends—partners in an extraordinary journey.

In July when I had finally recovered from my mother's harsh response to my phone call, I decided to call my father and ask to speak to Joud. He didn't seem to be surprised to hear from me but he asked no questions at all—nothing about how I was or where I was living, none of the concerns one would expect a father to have about his daughter, even a wayward one. Instead he said bluntly, "Joud is sleeping." That was the end of the call. I tried again a few months later because I heard from a friend that Joud, at age twelve, was being married off. My friend didn't know who Joud was marrying (I kept picturing some old man with my dear little sister), but she did say everyone was talking about it and presuming they wanted to get her married so that she couldn't escape as I did. My father only created excuses about why I could not speak to Joud and shared not a word about her. It made me think my friend was right about her impending marriage. But it also created a huge swell of nostalgia for me, thinking about the last time I saw all my sisters: Lamia wearing plain, dowdy clothing to please her husband, when she'd always found a way to add a bit of flair to the clothes we were expected to wear; Reem stumbling through her life as though she had left her potential in the mental hospital my father had taken her to, or lost it in the fog of drugs she was given; and now Joud, that little angel girl who loved me and looked up to me as her brave big sister. Was she being punished into a forced marriage because of me? These were painful thoughts that made me re-examine the steps I had taken. I wondered for a while if I should go back, endure the consequences I would suffer if only I could see my sisters one more time, if only Joud could be spared the fate that awaited.

In the fall of 2019, I phoned my oldest brother, Mutlaq, because again I was hungry for family news and hoped he'd give it to me. He was reasonably nice to me on the phone, asked a few questions about my new life and then, in a voice as hard as steel, he said, "Don't ever come back." I asked him if he had some information about what would happen to me if I returned, but all he said was "Forget about your family. Continue your new life in Canada."

I tried to reach out to him again in February 2020, again hoping to hear news about Joud, Nourah Mom and the rest of the family, but our conversation twisted its way into other issues that I wanted to share—issues he clearly didn't want to know about. I told him that I'd been raped in the back seat of a taxi while at university and that the beast who did that to me knew he would get away with it because I would be found guilty of losing the family's honour, and my life would be the price paid to restore that honour. I asked him how any society could condone such barbarity. He responded, as men in Saudi Arabia do, by saying if I had stayed in the house it wouldn't have happened. So I decided to let him know one of the many secrets I had about the so-called safety of our home—his friend had been harassing me for months before I left and was trying to have sex with me while he was in our house as a visitor. It was apparently too much for my strict, religious fanatic brother to absorb. He didn't say one word. He disconnected the line and blocked my number so that I could never call him again.

Now I depend on social media in Saudi to search for information. If Nourah Mom died, it would be on social media. If Joud is married, I think I would see it there too. I found out in the early summer of 2020 that Mutlaq was married, although I don't know who the bride is, and I learned that Majed moved away to another city. That's all I know, but I have never stopped wondering about them and am drawn still to various apps in the kingdom and invariably see that I still make the news on social media in Saudi. The haters, as I call them, still post horrid stories about me—mostly untrue and almost always calling for my death by hanging. But there is also news that allows me to know how the protests are going. For example, at about the same time as I arrived in Canada, the Shura Council (the legislative council in Saudi Arabia that advises the king) banned marriage under the age of eighteen but with an exception: they allowed marriage between fifteen and eighteen as long as there was court approval. By 2020 the courts had banned outright any marriage under the age of eighteen. So there's been a step of progress. And I hoped that could mean Joud is safe.

But if I dared to think the kingdom was advancing the rights of women, I soon found out I was dead wrong. Much of what was being changed—the child marriage law, for example— was more about show than reality. The deputy Middle East director of Human Rights Watch posted a comment on the organization's blog: "Rahaf Mohammed's courageous quest for freedom has exposed anew an array of discriminatory practices and policies that disempower Saudi women and leave them vulnerable to abuse." And he said, "Saudi Crown Prince Muhammad bin Salman wants to be viewed as a women's rights reformer, but Rahaf showed just how laughably at odds this is from reality when the authorities try to hunt down fleeing women and torture women's rights activists in prison." Sadly, I knew that to be true. Human rights organizations had been reporting that some of the women activists who were arrested were tortured with electric shocks, whipped, sexually harassed and assaulted in prison.

There were new rules in the kingdom, all right, and I continue to follow them: women can open a business without a husband's permission, and a woman was appointed head of the stock exchange, and mothers can keep custody of the children after divorce. But as for women being free to travel abroad, register a divorce or a marriage, or apply for documents such as passports without a guardian overseeing every step of the way—that doesn't happen where I come from.

There are still perils and consequences to pay for the drastic steps I took in running away. The freedom I sought is still up for grabs when my name is posted all over the place and everyone recognizes my face. I still frequently feel the conflict between the new Rahaf and the old Rahaf. And I know that even eighteen months after my exit from that suffocating place, its obligations and customs still swirl around my thinking the way thunderstorms bump into view, dump their contents and blow away. But I also feel joyously free—that no one can stop me from doing anything.

My goal in writing this book is to alert the world about the facts of a girl's life in Saudi Arabia and, even more, to send a message of hope to all the women who have had experiences similar to mine. Many of them have traded in everything—family, familiarity and a level of security, albeit laced with abuse—for uncertainty, often poverty and potential hazards that include being rejected by your chosen country and even being deported. Those of us who have landed safely in other countries still depend on the network that launched our escape—the people who acted like family as we navigated the fear, the loneliness and the dilemmas of making a home elsewhere, the ones who still watch out for each other.

I vowed to use my newfound freedom to campaign for women's rights in Saudi Arabia, and to call for an end to the male guardianship system enforced by the regime. There's no doubt that the number of women fleeing from the Saudi administration and abuse will increase, especially since the system they have in place right now isn't always able to stop them. I'm sure that as I write these words there are women using a secret code to get online and find out how to break away from the misery they live with. I hope my story encourages them to be brave and find freedom. But I also hope it prompts a change to the laws in Saudi, and that rather than being one girl's story of escape, this book becomes a change agent at home.

Gaining my freedom meant losing my family. Living in Canada means being mostly safe but also knowing that my face is familiar and my story is known, so I'm not 100 percent secure—but who is? My journey has been rocky, but it has allowed me to grow and learn and fulfill my dreams. Would I ever go back? I considered that once, when I feared for my little sister, but not now—I have goals to graduate from university and dreams to become an actor and plans to help refugee women settle. That's what I want to achieve. I have what it takes to make a good life.

A Letter to My Sisters
Who Need to Escape
the Lives They Are Living

You are not alone. *My life resembles that of many girls and women around the world who have experienced injustice, abuse and the denial of the rights that women deserve. I was beaten, threatened, raped by a stranger and hunted down as though I was a criminal by my family and the Saudi Arabian government during my escape. Like so many girls caught in the cage of oppression, I suffered through a depression so severe I felt internally dead and wanted to commit suicide. My recovery meant finding the way forward to a life that I felt I deserved, a life that would fulfill my dreams. I left everything behind, my family who I love and all that was familiar to me, because I believe I deserve a better life than the one they were forcing me to live.*

You also deserve freedom. You have the right to say no. You have the right to say yes. Do not allow another person to define your rights. Follow your dreams and fight to change whatever is holding you back. I surrendered twice, thinking I had to give in to my family and give up my hope for freedom. But then I gathered my courage and fought again. I fought the awful rules and the disgraceful presumptions about girls and women and I fought my fear of resisting. You need to fight too. Your place is not at the kitchen stove and the bed. You are a precious human being with a future that relies on work and play and happiness.

For the rest of my life, there will be people who will curse me and call me nasty names because I chose my lifestyle myself and broke Arabic and Islamic restrictions. I have already faced a heap of criticism and hurtful comments because I decided to be "me" and not the obedient, silent, invisible woman they wanted me to be.

My advice to you is this: Believe in yourself, be brave; don't wait for someone to help you, to set you free, to make you happy. You can do this for yourself. As I write this letter to you, I realize that eighteen months have passed since I ran away. During that time, I failed, I made mistakes, but I got up again, I learned and matured. Today, I can say that I have never felt better psychologically and physically, and I've never felt so happy and safe. The fight for freedom was worth it.

Remember, nothing is impossible, nothing is too difficult to try. I am an example of this. I fought the government and even the authorities in a foreign airport; I fought my family and my tribe; I fought everyone who tried to block my path to freedom and here I am.

Many women have fled, many others have tried and failed to get out of the clutches of a government that defies international laws and human rights norms. I don't want to encourage you to put your lives at risk, which is what you do when you decide to run away. I'd rather you keep up the quarrel with the government, the guardianship law and the mutaween brutes at home. But if that doesn't work, I have one word for you—ESCAPE.

Sources

58 "opening the door to evil": Xanthe Ackerman and Christina Asquith, "Women Haven't Really Won in Saudi Arabia—Yet," *Time*, December 15, 2015, https://time.com/4149557/saudi-arabia-elections-women-vote/.

58 "The game of chess is a waste of time": Lizzie Dearden, "Saudi Arabia's Highest Islamic Cleric 'Bans' Chess, Claims Game Spreads 'Enmity and Hatred,'" *Independent*, January 21, 2016, https://www.independent.co.uk/news/world/middle-east/saudi-arabia-s-highest-cleric-bans-chess-and-claims-game-spreads-enmity-and-hatred-a6825426.html.

66–67 "a man married to a woman": Razak Ahmad and Robert Birsel (ed.), "Malaysia's 'Obedient Wives' Anger Rights Groups," Reuters, June 5, 2011, https://www.reuters.com/article/us-malaysia-women-idINTRE7540FL20110605.

67 "end of virginity": Sebastian Usher, "'End of Virginity' if Women Drive, Saudi Cleric Warns," BBC News, December 2, 2011, https://www.bbc.com/news/world-middle-east-16011926.

82 Once, on a television program: "'Crazy Women' Angered by Polygamy Are Sinners, Says Saudi Cleric," *The New Arab*, October 7, 2018, https://english.alaraby.co.uk/opinion/crazy -women-angered-by-polygamy-are-sinners-saudi-cleric.

103 Excerpts from Farag Foda's *The Absent Truth* are taken from pages 222 and 224 of the 3rd edition (Cairo: Dar Al-Fikr li-l-Dirasat, 1980).

103–104 Excerpts from Nawal El Saadawi's *The Hidden Face of Eve: Women in the Arab World* are taken from page 20 of the 2007 edition (London: Zed Books, 1980).

112 The petition itself received 14,682 signatures on Twitter: Mazin Sidahmed, "Thousands of Saudis Sign Petition to End Male Guardianship of Women," *The Guardian*, September 26, 2016, https://www.theguardian.com/world/2016/sep/26/saudi -arabia-protest-petition-end-guardianship-law-women.

138 "It's been said that King Khalid offered a US$11 million bribe": Constance L. Hays, "Mohammed of Saudi Arabia Dies: Warrior and King Maker Was 80," *New York Times*, November 26, 1988, https://www.nytimes.com/1988/11/26 /obituaries/mohammed-of-saudi-arabia-dies-warrior-and -king-maker-was-80.html.

144 "To me, liberalism means simply to live and let live": Tristan Hopper, "What Did Raif Badawi Write to Get Saudi Arabia So Angry?" *National Post*, August 16, 2018, https://nationalpost.com/news/canada/what-did-raif-badawi-write-to-get-saudi-arabia-so-angry.

146 "I've encouraged women to go out": Princess Reema bint Bandar's statement is taken from Alicia Buller, "It's Time to Focus on Saudi Women's Capabilities, Not Their Clothes," *Arab News*, March 9, 2018, https://www.arabnews.com/node/1262466/saudi-arabia.

173 She posted a message on the internet that read: Dina Ali's post is taken from *Four Corners*, "Escape from Saudi," reported by Sophie McNeill and presented by Sarah Ferguson, ABC News, aired February 4, 2019, on ABC News, https://www.abc.net.au/4corners/escape-from-saudi/10778838.

184 "Rahaf's father is a governor in #Saudi Arabia": Mona Eltahawy (@monaeltahawy), Twitter, January 7, 2019, https://twitter.com/monaeltahawy/status/1082198562274971648.

190 "extremely worried that [a] Saudi woman": Phil Robertson (@Reaproy), Twitter, January 6, 2019, https://twitter.com/Reaproy/status/1081811017007394823.

191–92 "There are guards outside the hotel room": Sophie McNeill, "When a Saudi Teen Fled the Repressive Regime, Journalist Sophie McNeill Stood by Her," *Australian Women's*

Weekly, March 26, 2020, https://www.pressreader.com/australia/the-australian-womens-weekly/20200326/281595242607826.

197 **"UNHCR has been following developments closely"**: "UNHCR Statement on the Situation of Rahaf Mohammed Al-qunun at Bangkok Airport," UNHCR, January 7, 2019, https://www.unhcr.org/news/press/2019/1/5c333ed74/unhcr-statement-situation-rahaf-mohammed-al-qunun-bangkok-airport.html.

197 **"Don't believe it. Wait for the UN"**: Sophie McNeill, "When a Saudi Teen Fled the Repressive Regime, Journalist Sophie McNeill Stood by Her," *Australian Women's Weekly*, March 26, 2020, https://www.pressreader.com/australia/the-australian-womens-weekly/20200326/281595242607826.

198–99 **"Dear Rahaf, my @refugee colleagues"**: MME and agencies, *Middle East Eye*, "Saudi Teen Fleeing Family Now 'In a Secure Place,' UN Refugee Agency Says," January 7, 2019, https://www.middleeasteye.net/news/saudi-teen-fleeing-family-now-secure-place-un-refugee-agency-says.

200–201 **"Thai authorities have granted UNHCR access to Saudi national, Rahaf Mohammed Al-qunun"**: UNHCR, Twitter post, January 7, 2019, 8:05 a.m., https://twitter.com/refugees/status/1082261885683142661.

203 **"It could take several days to process the case and determine next steps"**: "UNHCR Investigates Rahaf al-Qunun's Case for Asylum," Al Jazeera, January 8, 2019, https://www

.aljazeera.com/news/2019/1/8/unhcr-investigates-rahaf-al
-qununs-case-for-asylum.

203 "The Kingdom of Saudi Arabia has not asked for her
extradition": Ibid.

203 "come forward and offer protection for this young
woman"; "deeply concerning"; and "It . . . comes at a time
when Riyadh is facing mounting pressure": Ibid.

203 "unable or unwilling to do so" and "Now that Ms
Mohammed al-Qunun has been given this status, another
country must agree to take her in": "Rahaf al-Qunun: UN
'Considers Saudi Woman a Refugee,'" *BBC News*, January 9,
2019, https://www.bbc.com/news/world-australia-46806485.

204 "She said very clearly that she has suffered both
physical and psychological abuse": Patpicha Tanakasempipat
and Panu Wongcha-um, "#SafeRahaf: Activists' Lightning
Campaign Made Saudi Teen's Flight a Global Cause," Reuters,
January 8, 2019, https://www.reuters.com/article/us-thailand
-saudi-campaign/saverahaf-activists-lightning-campaign-made
-saudi-teens-flight-a-global-cause-idUSKCN1P21CU.

204 "They should have [taken] her phone": Rahaf Mohammed
(@rahaf84427714), Twitter, January 8, 2019, 9:12 a.m., https://
twitter.com/rahaf84427714/status/1082641243690156032.
Original video was taken from @djboych9.

205 "There are, as I have just said, a number of steps": "Marise Payne Declines to Put Timeframe on Saudi Teen Rahaf Alqunun's Asylum Claim," ABC News, January 10, 2019, https://www.abc.net.au/news/2019-01-10/marise-payne-declines-to-put-timeframe-on-rahaf-alqunun-asylum/10706576.

206 "She needed to get out of Thailand very quickly": Robert Fife, "Saudi Refugee Who Fled to Thailand to Escape Family Is Headed to Canada," *Globe and Mail*, January 11, 2019, https://www.theglobeandmail.com/politics/article-saudi-refugee-in-thailand-headed-to-canada/.

208 "This is Rahaf al-Qunun, a very brave new Canadian": "Saudi Woman Arrives in Canada After Gaining Asylum," CityNews [video clip], January 12, 2019, https://toronto.citynews.ca/2019/01/12/saudi-woman-arriving-toronto/.

208 "It's obvious that the oppression of women": "'Brave New Canadian': Saudi Teen Rahaf al-Qunun Arrives in Canada," Reuters/Al Jazeera, January 13, 2019, https://www.aljazeera.com/news/2019/1/13/brave-new-canadian-saudi-teen-rahaf-al-qunun-arrives-in-canada.

213 "We are the family of Mohammed El Qanun": Sophie McNeill, "Rahaf al Qunun Pledges to Use Her Freedom to Campaign for Others after Being Granted Asylum in Canada," ABC News, January 14, 2019, https://www.abc.net.au/news/2019-01-15/rahaf-alqunun-speaks-first-time-from-canada-asylum/10716182.

229 "Rahaf Mohammed's courageous quest for freedom": "Saudi Arabia: 10 Reasons Why Women Flee," Human Rights Watch, January 30, 2019, https://www.hrw.org/news /2019/01/30/saudi-arabia-10-reasons-why-women-flee#.

About the Authors

RAHAF MOHAMMED was eighteen years old when she dramatically escaped from Saudi Arabia, capturing worldwide attention through her Twitter account. The daughter of a politician, Rahaf was raised according to an oppressive interpretation of Islam in which women and girls are given virtually no freedom. Thanks to a public plea for her life made on social media, she was eventually granted asylum in Canada, where she still resides, advocating for the freedom and empowerment of women.

SALLY ARMSTRONG is an award-winning author, journalist and human rights activist. She is the author of five bestselling books: the 2019 CBC Massey Lecture *Power Shift*, *Ascent of Women*, *The Nine Lives of Charlotte Taylor*, *Veiled Threat* and *Bitter Roots, Tender Shoots*. Armstrong was the first journalist to bring the story of the women of Afghanistan to the world and has also covered stories in conflict zones in Bosnia, Somalia, Rwanda, Iraq, South Sudan, Jordan and Israel. A four-time winner of the Amnesty International Canada Media Award, she holds ten honorary doctorates and is an Officer of the Order of Canada.

My Escape Route to Freedom

.... Land route
---- Air route

GREENLAND

CANADA

UNITED STATES

Toronto

ATLANTIC OCEAN

MEXICO

VENEZUELA

COLOMBIA

PERU

BRAZIL

BOLIVIA

PACIFIC OCEAN

ARGENTINA